Hammerstone

Hammerstone

A Biography of Hornby Island

Olivia Fletcher

Canadian Cataloguing in Publication Data

Fletcher, Olivia, (date)
Hammerstone

ISBN 1-896300-45-6

1. Hornby Island (B.C. : Island) 2. Geology—British Columbia—Hornby Island (Island)—History. 3. Coast Salish Indians—British Columbia—Hornby Island (Island)—History. I. Title.
FC3845.H67F54 2001 971.1'2 C00-911598-6 F1089.H63F54 2001

Editor for the press: Lynne Van Luven
Cover and interior images: Jane Wolsak
Author photo: Bob Cain
Cover and interior design: Ruth Linka

NeWest Press acknowledges the support of the Canada Council for the Arts and The Alberta Foundation for the Arts for our publishing program. We also acknowledge the financial support of the Government of Canada through the Book Publishing Industry Development Program (BPIDP) for our publishing activities.

NeWest Press
201-8540-109 Street
Edmonton, Alberta
T6G 1E6
t: (780) 432-9427
f: (780) 433-3179
www.newestpress.com

1 2 3 4 5 04 03 02 01 00

PRINTED AND BOUND IN CANADA

The hammerstone.

Table of Contents

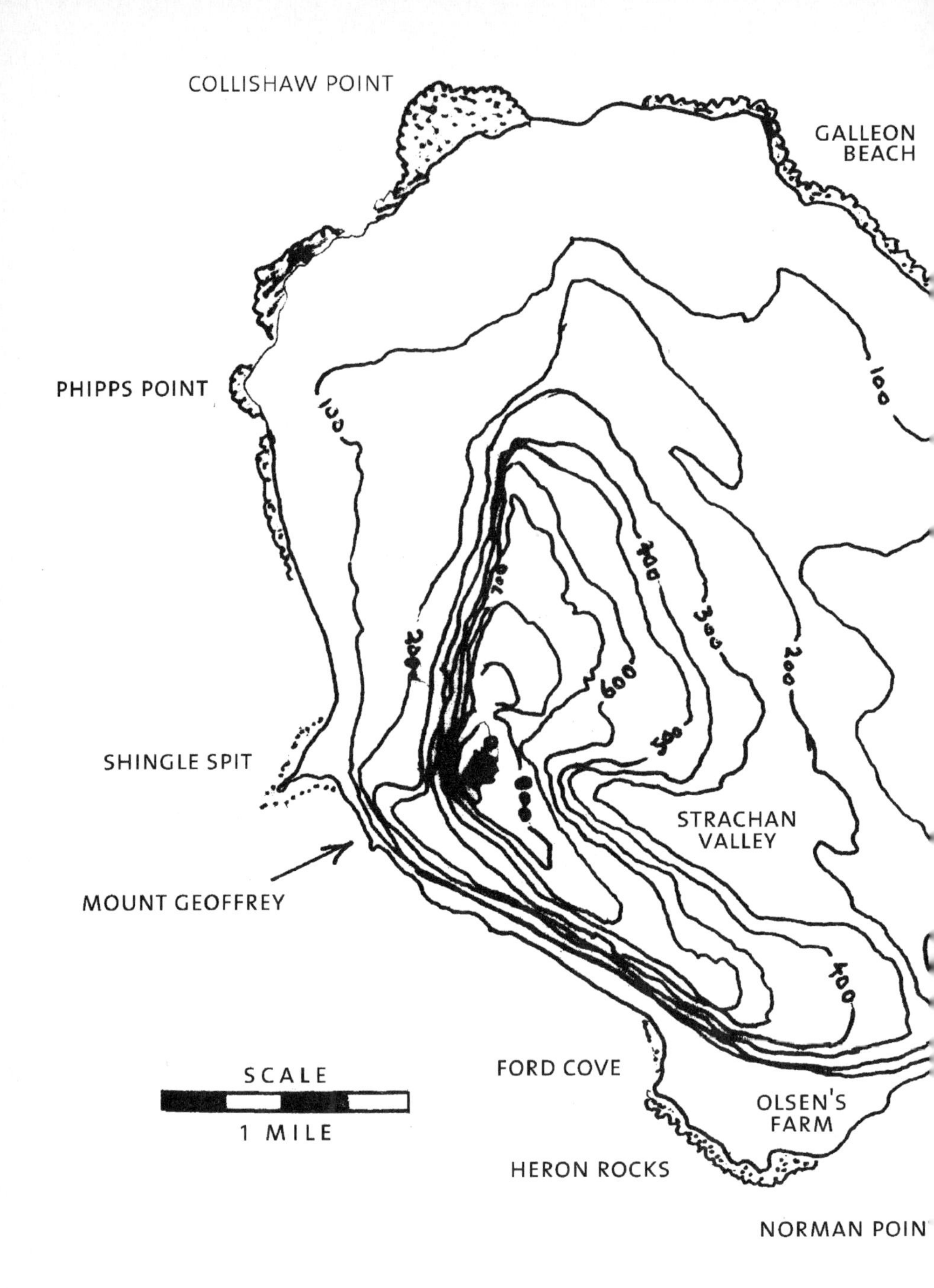

LAMBERT CHANNEL

HORNB

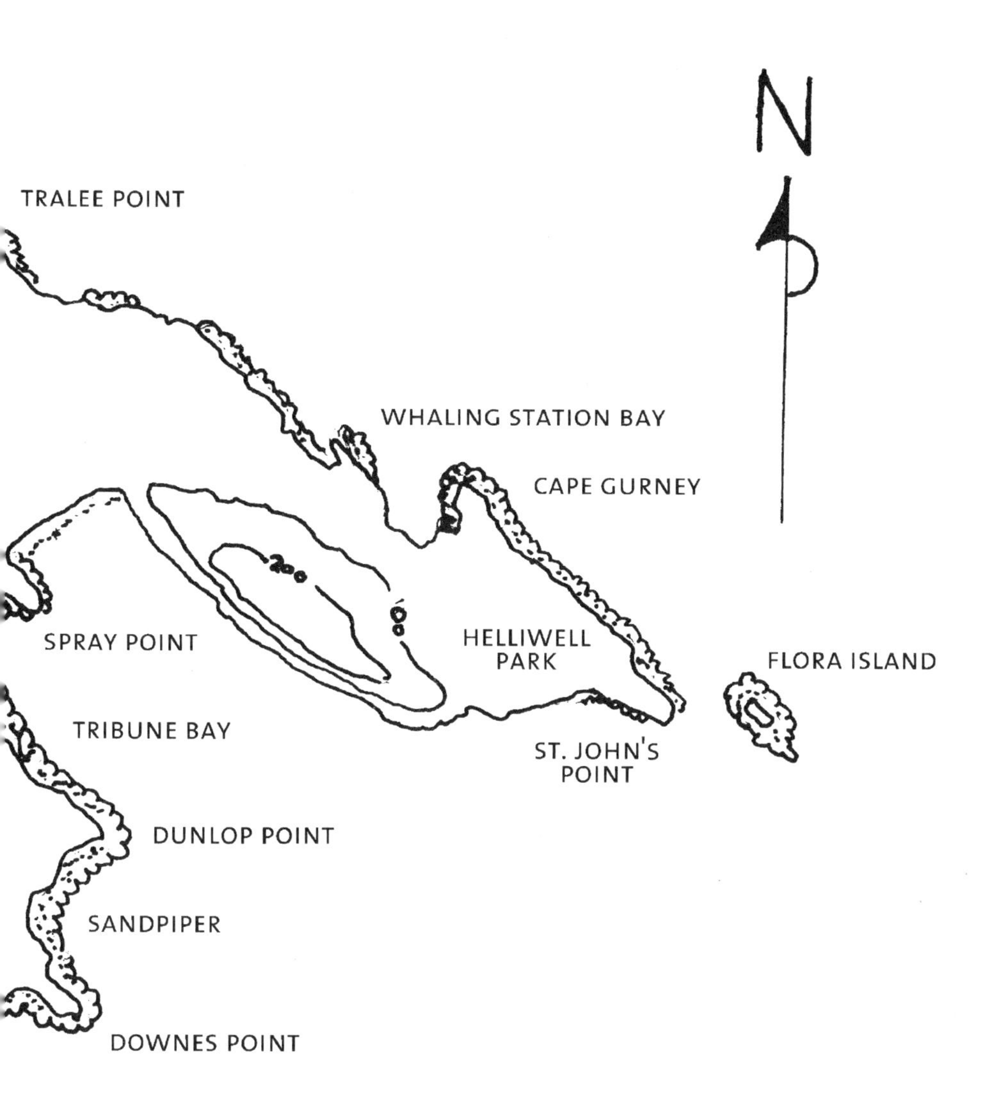
N
TRALEE POINT
WHALING STATION BAY
CAPE GURNEY
200
SPRAY POINT
HELLIWELL
PARK
FLORA ISLAND
TRIBUNE BAY
ST. JOHN'S
POINT
DUNLOP POINT
SANDPIPER
DOWNES POINT

SLAND

Foreword

WHEN HAMMERSTONE WAS FIRST PUBLISHED JUST OVER A decade ago, it was because I wanted to share with my fellow Hornby islanders what I had found out about the origins of the island. I had no idea when I started my research, seven years earlier, that my findings would eventually become a book. Since its publication in 1989, *Hammerstone* seems to have taken on a life of its own.

Newcomers and visitors will find the book a guide to the prehistory of Hornby. By picking up a pebble, viewing a glacial erratic, sifting garden soil through their fingers, they can, with a little imagination, take themselves back to other times. People living on one of the other Gulf Islands, or on the East coast of Vancouver Island, where the history both geologically and anthropologically mirrors the story of Hornby, may through this book experience the story of the place they call home.

"Goodness me! Whatever will they think of next!" exclaimed an esteemed elderly academic when I told her in the early 1980s about the movement of tectonic plates, and how the basement rock of Hornby had started in what is now the South Pacific. Now, such ideas about the movement of land masses are generally accepted.

The discovery in the late 1980s of an elasmosaurus skeleton on the Puntledge River at Courtenay, British Columbia, sparked an enthusiastic local interest in fossils and fossil hunting. This enthusiasm was bolstered in the 1990s by two books on Vancouver Island fossils (see bibliography).

Exciting discoveries have been made in the last twelve years. On Hornby, the first hesperonis fossil and beetle fossils have been found, more mosasaur bones uncovered, and in the layers of shale, a plethora of Cretaceous shark's teeth.

On Vancouver Island, dinosaur fossil bones have been revealed, giving paleontologists the first indication ever that dinosaurs did indeed roam and browse west of the Rockies.

Shadows of the Pentlatch, the Coast Salish people to whom Hornby once belonged, still show their presence on the land. On Hornby, as elsewhere in British Columbia, a meaningful understanding between today's aboriginal people and the newer settlers still remains to be achieved.

Hammerstone takes its readers backwards and forwards in time, helping them realize that evidence of everything that has happened in the past is still here. Stories are under our feet: in the rock formations; in the neglected middens, the petroglyphs, the fish traps, the still-plentiful food sources once used by the first people. And stories also linger in the relics of the early settlers, the stumps of enormous trees, cut down with double-ended hand saws, and in ancient orchards still bearing fruit. Even relatively new forest, where remnants of snake fencing show this was land once cleared and farmed, tells a tale to the observant visitor.

I have been asked to bring *Hammerstone* "up to date." Twelve years is perhaps no more than the flicker of an eyelid when the story being told covers 350 million years. The natural life of the island, with all its beauty and mystery, is still here. There are fewer fish in the sea, more evidence of sea lions and seals, and a new invasion of Canada geese. On the land itself, we see more sophisticated houses, tidier subdivisions and manicured roads, and fewer trees.

We Hornby islanders may have changed. Assisted in some ways, perhaps devastated in others, by technical innovations and economic globalization, the human net that covers the island has become more evident. However, Hornby is still a place of "direct experience." The computer, useful as it is, will never replace the joy we feel when we plod along a muddy path with our gumboots making deep suction sounds. Computer-mediated experience cannot replace the sweetness of cold damp wet leaves brushing our cheeks as we push through new trails, nor will it capture the quiet encounters that can take place between islanders, and the creatures—birds, deer, otters, seals—still sharing this island with us.

Introduction

THIS IS THE STORY OF HORNBY ISLAND, A SMALL ISLAND, no larger than twenty square miles, that lies, between Vancouver Island and the British Columbia mainland, in the Gulf of Georgia on the West Coast of Canada.

I started exploring Hornby Island's prehistory about twenty years ago hoping to find answers to some questions that had been tantalizing me: questions about the people who used to come to the island, about the ice ages and what was here before the ice ages, about the strange rock formations, and how the fossils fitted into the prehistoric scheme of things.

I thought I would find a factual progression. Instead, I found a piece of music, as though there were many harmonies and rhythms spiraling in constant play.

For Hornby, this music started 350 million years ago when one tectonic ocean plate, south of the equator, collided with another.

I have been led in my explorations from scientific papers into myth and the borders of consciousness and back again. During this journey, I have been touched by several different concepts of time: geological time, when everything is speeded up to enact drama; every-day time, when the buzzing of a bee or the fall of a pine cone can be heard; and seasonal and cyclical time, which is almost mythical time, or time tied to place.

With a Coast Salish hammerstone as the touchstone, the story of the island goes backwards and forwards, and round and round. What we can see and touch today—the boulders, sandstone, soil, trees, plants, fossils—are constants, again and again coming into view.

Although this book is about Hornby Island and its history, it could also be about any place or every place because at its heart lies the rhythm of the earth.

I have many people to thank for making this book possible. Perhaps living on an island was an advantage. Lacking library access, I was emboldened to seek my information from real-life sources. Dr. Jan Muller checked my initial geological writings. His encouragement meant much, as did Dr. Bruce Cameron's and Joanne Nelson's. Douglas Fiske sent me a copy of his thesis on the geology of Hornby Island. Dr. Glenn Rouse of the University of British Columbia loaned me papers normally not available to lay researchers, and Dr. J.G. Fyles his out-of-print book on surficial geology.

Katherine Capes, pioneer archaeologist in this area, became my friend, as did anthropologist Dr. Sarah Robinson, who schooled me in ethnology. Dr. Nancy Turner's approval and encouragement was great.

Among the native people, I had good talks with Mary Clifton of the Comox Band, and with the late Alfred Recalma of the Qualicum Band. Irene Seward of the Nanaimo Band, originally from Nanoose, read my book in draft, shared it with the Elders of the Nanaimo Band, and gave me her approval.

Among the writers who have helped me are Penn Kemp, the poet, and Sandy (Frances) Duncan, the novelist. The poets of West Word II, especially Barbara Findlay, helped weave the book's many threads into a coherent whole.

A thank you also to Bristol Foster, to Joyce Jeffries, and to Hornby Island poet Carole Chambers. And a big thank you to all the islanders who have helped me with their visions of Hornby Island today: Richard Martin, Pheobe Long, Eleanora Laffin, Katherine Ronan, and Tara Channel, to name only a few.

At the project's outset, Graham Beard of Qualicum encouraged me and guided me to relevant research material. Now, twelve years later, he has again helped me, and put me in touch with recent fossil finds. To him, special thanks.

And thank you Dianne Chisholm for telling NeWest about *Hammerstone* and to NeWest for deciding to publish it. To Ruth Linka and to Lynne Van Luven, many thanks! It is great working with you.

Finally, this book would never have happened if it was not for the patience and support of my family, particularly John. To them, and to all those whose brains I have picked through their writings, I say a deep thank you.

Prelude

WAVES ARE FOLDING OVER AND OVER, BREAKING AND licking the smooth mounds of rock, splashing against the dark seaweed-covered promontories. Our shoes are wet. We pick our way along a grassy path adjacent to the log fringed shoreline.

It is winter. You and I are walking on one of the northern beaches. Our hair is damp with rain, our cheeks pinched pink with cold. Oyster catchers, letting out shrill whistles, are alighting one by one on the dark seaweed, heads jerking, long bills probing. A small family of harlequin ducks are bucketing the wind-tossed waves.

High above our heads through wind-shredded clouds, we see two bald eagles soaring and gliding. Far below them is Hornby Island. What does it look like to them? A green moon snail with foot extended? An island encircled by small seas and other islands; an eagle's short flight over mountain ranges to the vast Pacific Ocean; an even shorter flight to the Coast Mountains, to jagged inlets thrusting into the vast hinterland of North America?

We walk along the beach to see some petroglyphs, carvings in the sandstone, which lie between us and the incoming tide. Clearly outlined are two killer whales, pecked and ground, pecked and ground again, many times.

The sandstone is covered with green algae at this time of year, so we are careful not to slip. The carvings once represented spiritual power. Now, in summer, they attract parties of school children, and people who like to take rubbings. Beyond the carvings, presently beneath the sea, is the scalloped shape of an ancient fish trap.

The log cabin, where twelve years ago we had tea and examined the hammerstone, is now rented to a young woman and her eight-year-old daughter. We settle ourselves outside the cabin on a log, and I take the hammerstone out of my pocket and hand it to you. It was found nearly twenty years ago on this beach, just west of Tralee Point. The hammerstone fits snugly into the palm of your hand, heavy but comfortable, an ordinary cobble except for a hollow on one side. Someone's hand, perhaps darker than ours, but no larger, has used this stone again and again as a hammer. Was it a favourite tool, mislaid after a camping trip to the north shore of the island by one of the first people?

It was here, being used, when canoes grated on the sandstone shore, when campfires glowed in the dark, when the chanting of song brought out the spirit within a carving.

And before this, for tens of millions of years, it was here as an ordinary cobble. A small lump of lava, its story is a thread in the story of the island.

PART ONE

The Travelling Land

A beaver. One of Hornby Island's residents.

Chapter One

TWENTY-FIRST CENTURY HORNBY

THE FEEL OF THE ISLAND DIFFERS FROM SUMMER *to winter. In winter, the beaches are slippery, the sun shines on the bare trunks of maple trees, and there is moisture in the air. The roads are emptier.*

Now it is summer. You are again with me. The grass has turned yellow, dust hangs along the sides of the roads, the water is warm. Hourly, the thirty-two car ferry from Denman Island spills forth visitors. The island population has risen to 6,000 from 850. Rental cabins, summer cottages and campsites are filled; strangers throng the Co-op corner; cars are parked this way and that. People stroll unconcernedly four abreast along the main roads, and the sandy beaches hum with the noise of radios, the cries of paddlers and swimmers, the splash of people in boats.

We islanders have almost lost sight of each other, and perhaps too of the real Hornby Island. Concerned with our own survival needs, we are organizing entertainments, growing vegetables, freezing fruit, making and selling handicrafts and food, catering to holiday makers, and in between all this, trying to give time to our own visiting families and friends.

Hornby, three ferry rides and five hours away from Vancouver, is a small island encompassing thirty square kilometres. It takes a day, when the tides are right in summer, to walk around it.

Within the sea perimeter of bluffs and beaches—some sandy, some pebbly, some sandstone with strange natural sculptures—are stretches of parkland, acres of woods and second-growth forests, fields in which rose bushes are scattered and through which sheep and lambs graze in ancient orchards. In clearings in woods there are homes, some spacious, some quite modest, each one reflecting someone's dream. The shoreline is still dotted with summer cottages, but larger, architect-designed houses are slowly taking their place. In the three subdivisions, houses cluster and tidiness is creeping in.

There are two stores. The Co-op store near Tribune Bay, with its gas pump and Ringside Market, is the largest, but there is also a small store at the Ford's Cove Marina. At the Shingle Spit, where the ferry docks, there is a pub and a restaurant, and a Credit Union with a banking machine.

The road from the ferry landing to Ford's Cove forms a large horseshoe and covers several miles. Sometimes it follows the shore line, giving vistas of sea; sometimes it passes fields, sometimes both sides of the road are thickly treed with underbrush. Indentations reveal, here and there, entrances to driveways. Signs advise visitors as to what homes or farms are open, what food or handicrafts may be for sale. Roads are more tailored than they were twelve years ago. Grass edges are neatly mowed; trees parallel to hydro and telephone lines have their overhanging branches pruned to prevent power outages. The feeling, however, as we journey around the island, is that Hornby is still rural, still what we believe to be Hornby.

Driving up the hill from the ferry, we pass on the right the Shire, once a commune, still flourishing with artists, writers, musicians. In among the remnants of what was once a farm are houses created by the owners, hidden in gardens with flowers, screened by young trees, grass thickets, blackberry and rose bushes.

A few kilometres on, we pass a large new campsite, a private airfield, a sheep farm selling old roses, a garden nursery, an organic vegetable farm, a cafe and bakery. Further along the road we see the recycling depot, the firehall, the Joe King Park (sports centre and social gathering place), the department of highways' yard and the Community Hall. Down a road to the left is the medical clinic, the festival and arts council offices, the RCMP offices, the preschool, the elementary school, the community school, the computer access centre, the health centre, the New Horizons centre, the library. . . . Like an ever-thickening net over its natural life, daily busyness covers the island. Notice boards, with handwritten as well as computer-printed messages, flood across an outside wall of the community hall.

At times over the past twelve years, it seemed that Hornby was becoming totally prescriptive, more managed, more fenced in, more money-oriented. But recently, there have been signs that the vitality and creativity for which Hornby is known is still very much here, still rooted in the island itself. Workshops in such subjects as wood turning or playing the blues have brought people to the island from all over the continent. The organization and initiative for these comes entirely from the islanders themselves. There is still much sharing, and should someone have a personal disaster, islanders still rally.

Beneath all this busyness, this temporal network of things happening, another equally vital island life carries on. This life involves harlequin ducks returning every year for their courtship ceremonies, owls hooting at night, mink scurrying under driftwood logs on the beach. It has something to do with rhythm and recurring patterns, with the tides, the moods of the weather and the changing seasons. And it's connected to the growth, death and rebirth of everything, including the rocks and boulders, which (if we watch long enough) are as fluid as the ocean surrounding these shores. I am hoping we can explore this hidden life together.

In the southwest corner of the island there is a small mountain, Mount Geoffrey. It stands about 1,000 feet above sea level, and dips on its southern and western sides sharply down to the sea. The slopes down its northern and eastern shoulders are gentler. For many years, the upper reaches of the mountain were preserved as Crown land, wild land on which many of us roamed, thinking it was all ours. This was a misunderstanding. The bench, facing Lambert Channel, has apparently always been privately owned; it is now up for sale. The remaining land, rescued from a Provincial Government "Crown Asset Designation," is now partly regional park, and partly a renamed, rezoned, "Groundwater Recharge Area/Sustainable Ecosystem Management Area." Dedicated islanders are protecting this land, making maps, improving trails.

We plod up one of these trails, an old logging road, past second-growth timber. Pausing almost at the top, we look down on the waters of Lambert Channel. The ferry, like a small toy, is furrowing its way across to Denman Island.

The summit of the mountain is a spacious mound of conglomerate rock covered with green moss, and surrounded by fir and alder trees. In the 1950s, after the last serious logging had taken place, there would have been a magnificent view from here: Southeast and below, Strachan Valley, and beyond, across the waters of Tribune Bay, the bluffs of St. John's Point Peninsular. From here on a clear day, it would have been possible to see up and down the Gulf for over a hundred miles. To the north, the mainland Coast Range mountains; to the southwest, Vancouver Island's Beaufort Range.

I take the hammerstone out of my pocket. It is black, flecked with grey intrusions. At one time it may have been a cobble in the conglomerate rock on which we stand, at the top of Mount Geoffrey.

Wisps of white mist are weaving among the treetops below us. It is time to go back. Back to where? To the busy world of Hornby? Or . . . back further into the unseen world? Back to Hornby's earliest beginnings?

The mist thickens. We are moving swiftly. In the same way that the past is with us now, in what we call the present, so will the present be with us as we journey through the past.

It is ninety years ago: we see Jessie French from Strachan Valley, granddaughter of one of the earliest settlers, exploring the unlogged sunlit forest, snaring grouse with her brothers.

Back a thousand years: we are aware of three men hunting deer, two women gathering herbs.

One, two, three, four . . . nine thousand years back: we watch the trees dwindle as the climate changes, and the sea rises to lap around the shoulders of the mountain.

Ten, twenty, thirty, forty thousand . . . one hundred, two hundred thousand years back: we experience the going and coming of four glacier advances and retreats.

Ten, twenty, thirty, forty, fifty million years back: the cracks and drops in the island's contours fashion themselves back into a rounded hill.

60 million years back: the hill is subsiding into an area of flattened rock.

It is 65 million years ago. We are no longer on a mountain, or even on an island; we are on an extensive delta. To the southwest, sandstone rocks, drifts of pebbles, sandbars, merge into a coastal fringe of palm trees, behind which rise green hills. The lower reaches are wooded with flowering trees and vines, and the upper are covered with coniferous forests of redwood and cypresses. Ferns carpet the ground. To the northeast, we see water, more water and islands.

We are also no longer just north of the forty-ninth parallel. While spinning backwards in time, we have moved south, to the latitude of present-day California. We are standing on the newly made sedimentary rocks of Hornby, on some of the rocks we know to see and touch.

We have not, however, reached the geological beginnings of Hornby. The pebbles in the conglomerate, the sand of the sandstone, and Hornby's basement rock, all started life millions of years before this. We have a far faster, possibly more violent, backwards journey to go.

We are still moving south: it is now 170 million years ago, and we are in the latitude of present-day Mexico. Massive volcanic eruptions are taking place. We move again through time.

It is 350 million years ago: we have reached the Panthallasah Sea. Just south of the equator, Hornby Island is being born.

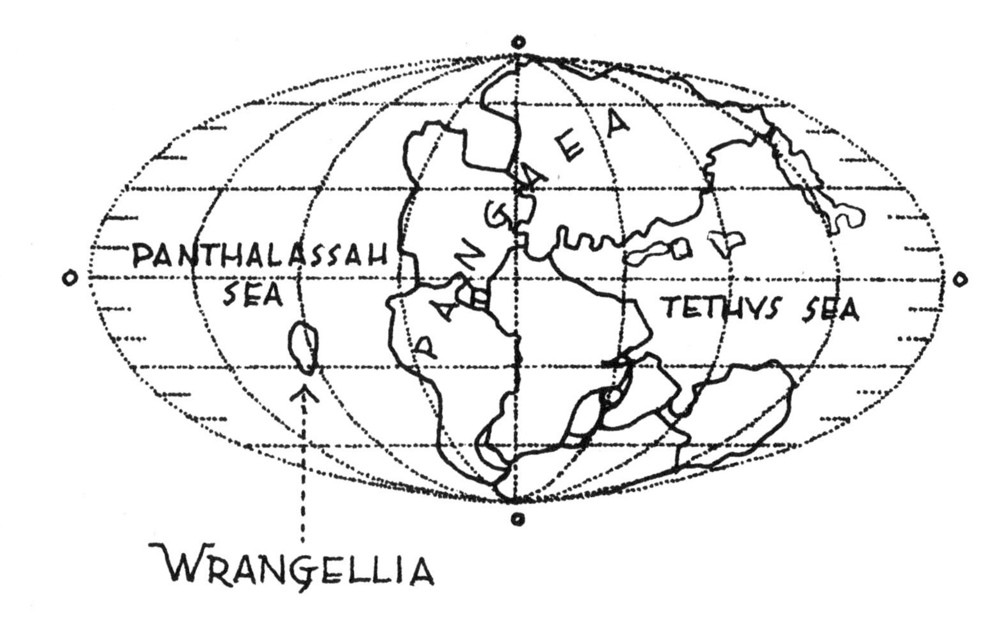

The continents as the universal landmass of Pangaea about 250 million years ago.

Chapter Two

350 TO 170 MILLION YEARS AGO

THERE IS TROUBLE IN THE PANTHALLASAH SEA. EXPLOSIONS are following explosions. Steam and dust, jettisoned high into the sky, are smudging out the sun. Tidal waves forty feet high are thundering across the surface of the sea. Rumblings, hissings, grumblings batter our ears. The world beneath us is a mass of noise and fermentation.

One ocean plate meeting another has rifted. The older plate, plunging beneath the younger, is melting. As it disappears back beneath the mantle of the earth, remnants of the plate together with new magma from the mantle are spurting hot lava miles high into the air.

It is 350 million years ago, and we are watching Hornby Island being born.

Along with Hornby, which at this point is little more than a hundred thousand tons of bubbling basalt, a volcanic arc is being created. This volcanic arc is responsible for creating almost the whole of Vancouver Island, the Queen Charlotte Islands, part of the mainland and part of Alaska. Geologists call it Wrangellia.

The globe below us looks quite unlike today's map of the world. All the continents, Eurasia, the Americas, Africa, have been pushed by moving oceanic plates into one great land mass, Pangaea. The continents, or continental plates as we should call them, have been detached from each other before this, and will separate again. They are experiencing temporarily, for a period perhaps of several hundred million years, a kind of traffic jam.

Wondering what it was like to be in the world 350 million years ago, we drop down into Pangaea and find ourselves on the fringe of a steamy green forest. Across a slimy swamp, to our right, in and out of gigantic ferns, slithers a scaly amphibian. A dragonfly, its wing span over three feet, flutters across the swamp into the forest. We follow. Giant club mosses and horsetail trees stretch up forty to sixty feet to a distant sky, their trunks wound with strange climbing plants. There is no bird song, no colour from flowers. Instead, we hear the clack of insect wings and the sound of dripping water. The ground is gold from pollen dust.

The horsetail trees towering above our heads are hardly recognizable as ancestors of the waist-high weed that commonly grows today on Hornby, the plant with jointed stems that can be pulled apart and put together again.

Escaping this fetid world of gold and green, we look again at Wrangellia. It is now a large expanse of dark grey volcanic rock on the surface of the ocean. We hear an occasional explosion as more molten lava comes to the surface.

How is it known that Hornby Island originated in the South Pacific? We are temporarily back in the twenty-first century, sitting on a log looking across at Texada Island.

As volcanic lava cools and turns into basalt rock, frozen in the hardening rock are molecules of iron compounds which, through magnetic attraction, align with the North Pole. The angle enables scientists to establish in what latitude eruptions take place. Unless the poles had switched places at this time, which they sometimes do, Wrangellia surfaced south of the equator.

The name given by geologists to the earliest rock of Wrangellia is the "Sicker Formation." The bottom layer of Hornby Island, which we cannot see since it is beneath the sedimentary rock, is part of this formation. Some of the pebbles on the Hornby beaches, much metamorphized, eroded, polished, tumbled and washed, date back to this time.

Picking up handfuls of colourful pebbles and dribbling them through our fingers, we wonder. Some of the grains in the sandstone also originate with Hornby's earliest beginnings.

Fossils reveal the approximate date the early Wrangellian eruptions took place. In some of the Buttle Lake limestone, deposited many thousand years later on Sicker rocks, fossilized brachiopods, bryozoans, foraminifers, pelecypods, corals, gastropods and ostracods have been found. Through their distinctive shapes and patterns, these sea creatures tell paleontologists when they lived.

Over the Gulf of Georgia from Hornby Island, in front of the snow-capped Coast Range, rests Texada Island. Across its greenness there is a slash of white where limestone is being mined. Some of Texada limestone dates back to early Wrangellia. In later Texada limestone, early ammonite fossils have been found—but this is jumping ahead in our journey through time.

Back in geological time, the seas are quietening, and the bare rocks of Wrangellia are turning green with lichen and ferns. The sun shines on rippling white water. Through erosion and changing sea levels, Wrangellia is slowly sinking.

More thousands of years pass. Wrangellia is now under the water being carried north on the back of the moving Pacific Ocean plate.

Under the sea, ray-finned fish and sharks are swimming in and out of submerged mountain ranges, snapping at free-swimming shelled creatures. A million million organisms—brachiopods, bryozoans, foraminifers, pelecypods, corals, gastropods and ostracods—are drifting down into the underwater valleys, softly piling up, slowly making limestone.

It is 170 million years ago. From high above the globe, we see the supercontinent Pangaea slowly breaking up, becoming again several separate continents.

To the northeast, the North American plate is travelling west. Its western edge ends with the Province of Alberta. There is as yet no proto-Rocky Mountain Range, no embryonic British Columbia. On the other side of the North American plate, the North Atlantic is opening up. The African plate is still attached to the South American plate.

Wrangellia, far from land and still under the water, meets a "hot spot," a plume of hot rock welling up from deep in the earth's mantle. The massive Sicker rocks are shaken by violent, continuous earthquakes and volcanic eruptions. We see lava flowing over the ancient Wrangellian rocks and over the early limestone—first underwater, then above. Molten lava is cooling into rock, and what is known as the Karmutsen Formation is being created.

Wrangellia comes to the surface.

Chapter Three

170 TO 80 MILLION YEARS AGO

WE ARE IN THE LATITUDE OF PRESENT-DAY MEXICO, STILL some distance from the North American plate. On the North American plate, early dinosaurs are browsing on ginkgo and cycad trees. On its higher reaches, ferns and early conifers grow.

Wrangellia also had cycad trees. Fossilized cycad leaves dating to this time are known to have been lying in Karmutsen Rock on the Queen Charlotte Islands. In the last twelve years, they have also been found near Nanaimo.

We have slowed down into everyday time. Between the occasional lava flows, the landslides, the changing sea levels, and the fracturing of rock, we watch soil being built up, and mosses and lichens, followed by plants and trees, being established.

Time is now ticking at a pace we understand. We land on Wrangellia, feel the warmth of rocks, dip our hands into streams gurgling towards the sea, listen to the plop of falling cycad cones. Night comes, bringing a full moon tracing a path of silver on an endlessly rippling sea. The air is warm. The night passes. The moon fades, there is a smudge of pink in the east. A tip of orange light grows into an orange ball. The sea is bathed in colour.

The world lightens. A marine creature pulls itself ashore. We leave.

Speeding up into geological time, we watch the soil and vegetation being swept away. Wrangellia is again sinking beneath enveloping seas.

Back on the beach in the twenty-first century, we again look at the hammerstone. It is a lump of amygdaloidal lava—lava that solidified into rock above water. There are deposits of amygdaloidal lava in the Karmutsen Formation in the Beaufort Range. Many of the cobbles and pebbles on Hornby date to this time.

130 million years ago. Far away to the northeast of Wrangellia—now a string of islands slowly moving north—we can see the North American plate. The sea sweeps south of the North American plate and round into the expanding Atlantic.

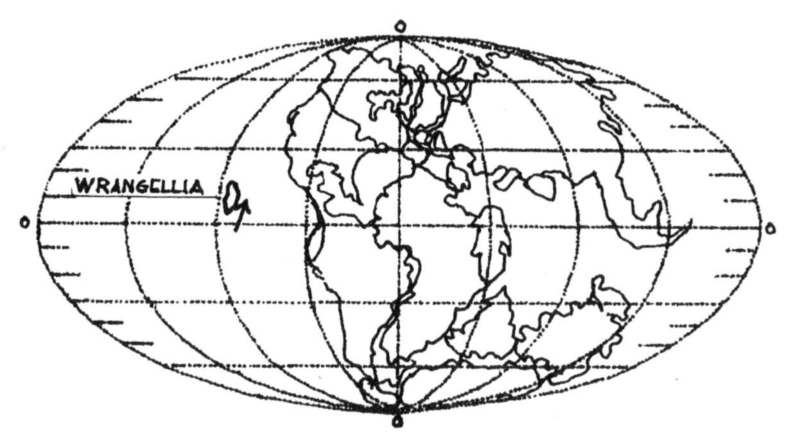

The Jurassic Age, 130 million years ago.

To the north, and between Wrangellia and the North American Plate, are many different terranes, land masses similar to Wrangellia, but of differing sizes, origins and ages. They are twisting, turning and moving at different speeds, like boats docked but not tied up.

Unlike the continental plates, which are pushed by oceanic plates, independent terranes, such as Wrangellia, travel on the top of oceanic plates. They are rather like soap suds on the top of water. When the oceanic plate plunges back into the earth's mantle, the independent terranes remain on top.

Volcanoes start up on the western edge of the Wrangellian islands (today the west coast of Vancouver Island). Steam and ash and molten lava are being jettisoned into the air. The Pacific Ocean plate has created an underwater trench as it plunges back into the earth's mantle, melting some of its existing rock as it goes and releasing great heat and magma from the earth's core.

New mountains are being created. Through heat and pressure, ancient sedimentary rock is being transformed into metamorphic. Beneath existing rock, new lava, injected from the earth's centre, creates granite batholiths. Old mountains are crumbling. The "stuff" of many of the pebbles on our beaches—porphyry, jasper, chert, agate, argillite, schist, breccia—is being made.

Some of the Wrangellian islands subside beneath the sea. The volcanic eruptions quieten.

To the north, Wrangellia is slowly joining another detached piece of land, the Alexander Terrane (today seen as part of the Queen Charlotte Islands and the adjacent mainland). Then to the east, Wrangellia docks along side the Stikine Terrane, another piece of independent land form, today recognizable in a stretch of the mainland bordering the western seas.

Plate action has shifted from the outlying islands of Wrangellia to the long narrow stretch of sea (now known as the Gulf of Georgia) between Wrangellia and the Stikine Terrane. Tongues of fire lash upwards, with swirls of smoke. Rocks explode. Clouds envelope and disperse. Hot lavas spurt and fall. There is a long pulsating gash from far north to south. The lands surrounding shudder, and tidal waves rush here and there.

We are watching the embryo rocks of the coast range being made.

Cretaceous sedimentary rock today.

Chapter Four

THE CRETACEOUS WORLD

WE ARE NOW IN THE CRETACEOUS AGE, THE GOLDEN PERIOD that covers 70 million years, and have reached the time when flowering trees and shrubs appear in the world, birds proliferate, and dinosaurs continue their dominant role as land vertebrates.

Over on the North American continental plate, divided in half by water—for a sea in the north has joined a sea from the south—there are browsing herds of horned dinosaurs, duckbill dinosaurs, and bonehead dinosaurs. In the swamps and lakes are crocodiles, lizards, snakes, turtles and frogs, and freshwater fishes. Hidden in the undergrowth are the first mammals, tiny insect-eating, opossum-like marsupials, and fruit eating multituberculates.

Until recently, the lack of fossil evidence seemed to indicate that there were never any dinosaurs west of the Rockies. However, in the last dozen years there have been discoveries of dinosaur fossils on Vancouver Island, and recent fossil finds in Nanaimo have given new authority to the existence of certain sea creatures that, up till now, we had only suspected to have swum in our Cretaceous waters. New research has further confirmed the presence at this time of forests of giant redwoods—palms, cycads, ferns, and flowering plants. Here, on Hornby, fossilised hesperornis bones have been found. A hesperonis was a large wingless diving bird, whose vigorous hind legs served as paddles, and which, unlike the other birds, still had teeth.

It is 80 million years ago. We are in every-day time above the waters that cover Hornby's basement rocks. The mountains of Wrangellia are behind us. Across to where we now see Texada and the Coast Mountains, stretching for hundreds of miles, are large and small islands, punctuated with steam and dust from grumbling volcanoes. The latitude is that of present-day California. The climate is semi-tropical.

A pterosaur, a flying creature with leather wings, launches itself from a cliff of Wrangellian rock, and circles on wind currents the stretch of water above the basement rocks of Hornby.

Beneath the water, the ancient eroded metamorphized rocks are encrusted with scallop-like and snail like creatures. Through waving sea-weeds, chiriocentrids, herring-like fish, swim in shoals.

On the surface, an elasmosaur, paddling with its four flippers, coils back its excessively long neck and strikes out like a snake at the pterosaur. An upwind carries the pterosaur out of its reach. The small head on the long neck then dips under the water and swallows a small fish.

Turning away from this scene of islands and waters, we look down on what is now the Trent River on the east coast of Vancouver Island. Forests are breathing a thousand different shades of green as oaks, walnuts, elms, birches, willows and poplars burgeon into leaf. Along the present Tsable River there are groves of koelrutia trees, golden with showers of flowers, and maple trees tassled yellow green. Above Cumberland, we see alders, cascaras, viburnums, sweet-scented shrubs and vines growing. And up in the mountains, under tall cypresses, conifers and redwoods, ferns of many kinds carpet the ground.

We are reaching the time when the rocks of Hornby, the ones we see and touch today, are laid down in layers of sediment on the island's Wrangellian basement rocks.

Still above the present-day east coast of Vancouver Island, but now in speeded-up geological time, we watch the mountains of Wrangellia eroding. Streams and rivers trickle, splurge, waterfall, roar down though woods, carrying and collecting as they flow, sand and pebbles, uprooted trees, small boulders.

On reaching the sea, the rivers of the Wrangellian islands drop their detritus, and a vast delta is formed.

Layer on layer of debris from the rivers is hardened, by chemical action and by pressure from further sediment into rock.

In between one layer of sediment being laid down and the next, we see soil being built up and ferns, trees and shrubs taking root. We then see these trees and shrubs being trapped in the next flow of sediment. And first peat, then coal, being made.

Back in the twenty-first century, we look at Bell's book on the Upper Cretaceous Flora. Here there are photographs of the imprints of leaves found fossilized in the Comox, Cumberland, and Nanaimo coal fields. When interpreted, these leaves give a glimpse of the world through which the sediment of Hornby will have flowed.

The fossilized leaves captured in the coal seams tell us that the ancestors of our familiar alders, birches, cascaras, walnuts and magnolias were here before the sedimentary rock of Hornby was formed.

Sensing, smelling, touching today's trees perhaps makes the trees of 65 to 80 million years ago seem not so far away? How dependable are our established concepts about the logical progression of time? A telescope many billions of miles away in space focusing on the planet Earth would see, because of the time it takes for light to travel, not cities and highways, but Cretaceous forests and browsing dinosaurs.

We watch rivers, as the mountains erode, spilling forth pebbles that will one day form conglomerate rock; spilling forth sand that will become sandstone; spilling forth eroded rock of all kinds that under the rising seas will turn to shale or mudstone. As the sea level rises and falls, it seems as if Wrangellia is being rocked. Sediment from the eroding hills is being laid down in a pattern that keeps repeating.

During this period of 15 million years we watch new land being created, from Port McNeil in the north to Galiano Island in the south. It is at the very end of the cyclical pattern of erosion and sedimentation that the sedimentary rocks of Hornby come into being.

Chapter Five

THE ISLAND'S SEDIMENTARY ROCK

THE SEDIMENTARY ROCK KNOWN AS HORNBY *Island was laid down in layers 70 to 65 million years ago, in the latitude of present-day California. Each layer came from different sources of eroded mountain side, was spilled out from rivers, and later hardened, under different physical conditions.*

Each cycle of sediment starts later, and stretches out further than the cycle of sediment preceding it, south-west to north-east, much like a pack of cards idly flung across a table. The first formation is the De Courcy. The Northumberland lies over this, then the Geoffrey lies over the Northumberland, which is followed by the Spray. The final cycle is the Gabriola, seen today as the upper level of rock on the St. John's Point Peninsular, the foot of the moonshell, part of which is now Helliwell Park.

The deep water of the Lambert Channel wallows under an overhanging rock, as we walk along the De Courcy sandstone shelf from Ford's Cove to Heron Rocks. The smoothness of the sandstone is pock-

marked, perhaps from rain when the sandstone was hardening, perhaps from sometime-burrowing crustaceans, perhaps from inconsistencies in the cementation. The criss-crossing of cracks and crevices are relics of earthquake tensions from plate movements, both now and back into 10,000 centuries of time.

A sand-impacted bank, on our right, towers up to trees and bulges out over layers of grey and orange shale. Every few feet, between the shale and the hardpan, nodules of hard concretionary rocks are embedded like beads in a necklace.

We find a loose one, and with a rock hammer break its hard spheroidal shape in half. There is nothing in it, just solid rock. It might have contained a fossil because the nodule will have been formed by a nucleus of rotting organic matter, a piece of shell or a dead animal. Organic nucleii, when decaying, create a porous place to which a concentration of minerals is attracted.

Sometimes iron oxide creates the hardening process, sometimes marcasite, or carbonate or silica, each of which, because the solubility of each mineral is not consistent, erodes differently.

Back in the Cretaceous, we are looking at the De Courcy Formation as it was 70 million years ago. All around is a vast delta, a mosaic of sand islands, criss-crossed with rivulets. Dead trees lie abandoned on sandbars, their limbs torn and scarred from journeys down rivers in spate.

Moving out of everyday into geological time: sand and sediment are being washed down the rivers, dead trees and rotting organic matter are being buried, rivulets are changing, and again changing course; banks of sand are being tumbled.

As lithifaction, the hardening of the sediment through chemical interaction, temperature and the weight of further sediment, takes place, much of this action will be preserved in the geological record.

Perhaps as much as any archaeological dig, though in a different time frame, Hornby beaches are packed layer on layer with history.

The honeycombed surfaces we see today in the De Courcy sandstone are produced by the weathering of a carbonate-filled network of cracks. The sea washes out the softer sand, leaving the sandstone, lithified with more cement, standing out in relief.

The swirls and dips and sculptured shapes on the Hornby beaches are largely due to these inconsistencies in the cementing processes. The waves of a million, million seas wash away the softer sandstone, capriciously shaping and fashioning whatever happens to be left.

It is a grey day. White cloud hides Vancouver Island, Denman Island is no more than a shadow across the channel. We are still in the twenty-first century, still on the southwest side of Hornby but are now between Phipps Point and the Shingle Spit, on the mudstone rock known as the Northumberland Formation. This is the flow of sediment that came after the De Courcy sandstone, and which was hardened under the sea. The beach is covered with a mosaic of fine gravel, seaweed, and shallow pools of water. At our feet there is a scattering of stones and small boulders, and also some dull looking large grey sandstone pieces. We tap one of them. It cracks open, revealing the rainbow colours of an ammonite: an ammonite over a foot in diameter. A solitary gull soars above the beach.

Our eyesight grows hazy, the bank with its trees is disappearing.

We are still standing on the sandstone of the Northumberland Formation, but the view in every direction is quite different from what we see today. There is no Denman Island, no seaweed as we recognize it, no scattering of small boulders and rocks; to the northeast there is no Hornby Island, just a flat wetness.

The composition of the sandstone and the texture of this Cretaceous beach is similar to that of today. However, today's beach is many, many feet below. The twentieth century beach, because of erosion over millions of years, will represent an earlier flow of sediment. In examining layers of geological history, we go backwards in time.

The sea under which the Northumberland sediment was deposited has already receded. In the wake of the subsiding sea, the Cretaceous rivers, still carrying a little sediment, have created mudflats, small islands and marshes. On the shoreline, heron-like birds are fishing. Over sunlit water, ichthyornis, birds similar to terns, soar and swoop, while baptornis, grebe-like birds, paddle along in search of fish. They are careful to avoid the hespornis, the large swimming bird with teeth. The sea is deep, warm and rich in marine algae.

Gulls cry, the water laps on the sandstone beach, and the wind gently swishes in the flat leaves of the palms. There are earth-tremors. It starts to rain. Down the rivers pebbles start to tumble. The next sedementation of Hornby, the Geoffrey, is starting.

The Cretaceous period was once thought to be a time of geological quiet. Now it is being discovered that there was tectonic movement and mountain building. Plate movement created the Coast

Mountain range at the beginning of the era. And plate movement would explain the changing sea levels, and the erosion of the Wrangellian mountains.

Earthquakes as well as torrential rain were needed to move the boulders and massive rocks found today in the Geoffrey conglomerate.

Down the overflowing gullies tumble torn-up trees, pebbles, cobbles and boulders, some up to two feet in diameter. Down over the Northumberland formation sweep pieces of sandstone from earlier sedimentation, pieces of gneisses, schists, chert, argillite—rocks of all kinds from the crumbling Sicker and Karmutsen Formations. In among the moving debris are fossil invertebrates in concretionary cobbles, relics of a different time.

There are lulls in the bombardment. At times fine sediment sifts in to form layers of sand between the stretches of layered rocks and boulders. The swollen rivers drop the heaviest cobbles first. The smallest are carried north, to what is now Galleon Beach.

This vast expanse of pebble beach eventually lies quiet and begins to lithify into conglomerate. 15 million years hence it will have been folded into a hill and then, after 25 million years, earth tensions will create first one and then two faults; the cliffs of Mount Geoffrey will slowly, slowly be cracked and dropped into place.

We are back again on the top of Mount Geoffrey. It is a beautiful, late-summer day in the present. We kneel down and remove some of the moss. Beneath a thin layer of peaty soil we can feel the Geoffrey conglomerate. In these pebbles is evidence for much of the evolution of Wrangellia.

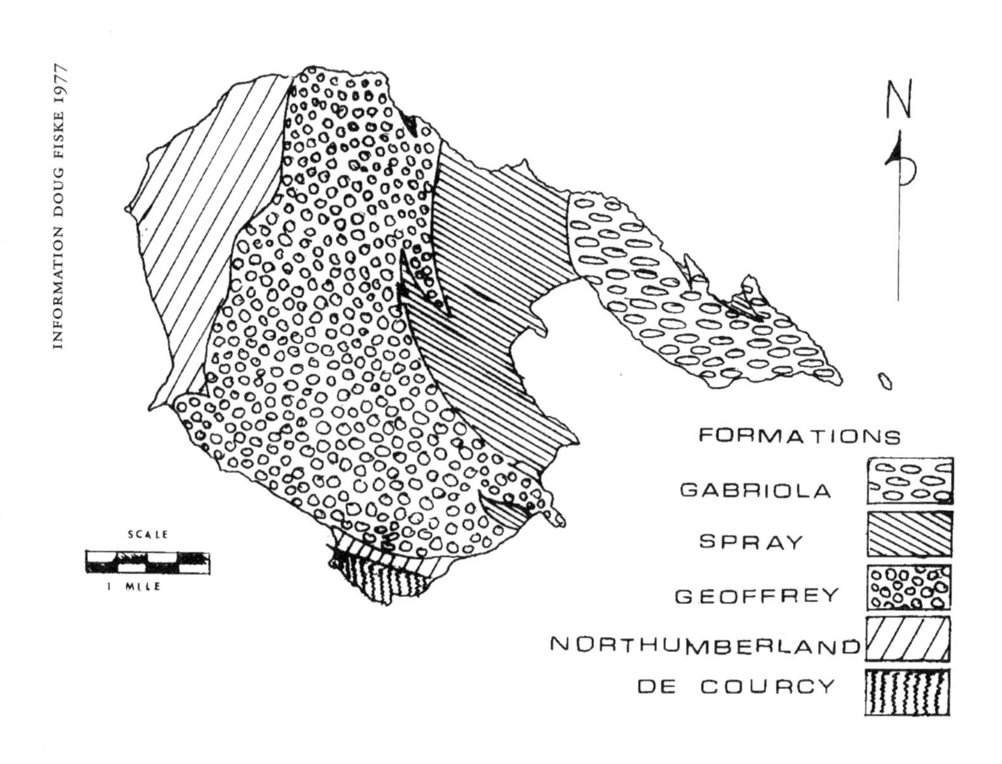

Simplified map showing roughly where the different cycles of sedementary rock laid down 70 to 65 million years ago may be seen today.

Chapter Six

FOSSILS

THE TIDE IS LOW. WEARING STOUT SHOES, WE *scramble down the cliffs on the western boundary of Helliwell Park on the St. John's Point Peninsular and clamber along the almost inaccessible piece of beach between the Park and Tribune Bay. There is about a mile to go. Some of the rock here is conglomerate from the next formation, the Gabriola, but most of what we see—eroded, battered and slumped, and at some time faulted—is the cemented sediment following the Geoffrey, named by geologists the Spray Formation.*

From this formation also come the sandstone sculptures and the sandstone spur of Spray Point, the undulating beaches on the north shore, the rock underlying the conglomerate of the St. John's Point Peninsular, the papery shale of Sandpiper, and the mudstone of Fossil Beach.

The sand that created it was cemented into rock, partly in a delta environment, similar to the De Courcy Formation, and partly in a shallow sea environment similar to, but about 2 million years later than, the Northumberland. We are looking at rocks created 66 million years ago.

We make our way along the narrow passage of rock between the two parks. The sea laps on our left side, and cliffs tower on our right. We pass sandstone cannonballs, two feet in diameter, concretions of all shapes and sizes, boulders of every size and kind. Some boulders are so exquisitely sculptured and patterned that, placed singly in an urban setting, art critics would write about them. Others, made of conglomerate, seem to be studded with semi-precious stones.

Towering above us to the right are yellow sandstone cliffs reminiscent of crumbling walls in Mediterranean ruins. Further along, the cliffs have gargoyles of harder rocks extruding. There are caves, and ledges on which wild flowers are growing. Shy pigeon guillemots, knowing this area to be almost inaccessible to humans, are resting on some of the rocks. Their enormous, red webbed feet are splayed out over the dark boulders.

As we near the sands of Tribune Bay there are, on our left, strange angular rocks set in orange sandstone. A geologist says they tell the story of a partly hardened channel, part of a delta in the Cretaceous, through which the sediment flowed. Earth tremors cracked and dislodged parts of the channel's bank, creating semi-lithified broken pieces. These later became reset in a matrix of softer sand. The "orange stuff" is the iron oxide that cemented the sand.

At Tribune Bay, a friend picks us up and drives across the island, past the Co-op Store, the community hall, the firehall, past woods and fields, to Savoie Road. She turns right. Half a mile down this gravel road she parks the car, and we walk the short trail to Fossil Beach, one of the northern beaches.

It is still low tide. The Spray Formation here looks like a moonscape. Vast stretches of shale and mudstone are blotched with pools of water. Zones of seaweed are broken by large boulders and rocks.

Several amateur paleontologists on the beach are searching for Cretaceous fossils. One man, wearing rubber boots, windjacket and woollen cap, is removing, flake by flake with a small knife, the shale from around a brown lump. A glint of mother-of-pearl confirms to him that it is a fossil. As he whittles, the conically shaped piece reveals itself as a baculite, the straight member of the usually spiral-shaped ammonite family.

Nearby, another man is easing the slim, long shape of a dentalium out of its ancient bed. Whack, someone else has broken open, with a special hammer, a hard grey lump of sandstone. There, in his hands, are the glossy curves of a coiled ammonite. As well as baculites, ammonites, and dentalium, spiny lobsters, pelecypods (fossils that look like large clams) and pieces of wood are found here.

This beach, near Collishaw Point, is the main source of fossils on Hornby. Graham Beard of Qualicum Beach picked up a mosasaur skull here some years ago. Mosasaur bones are commonly found on this point. Mosasaurs were marine lizards. They grew up to twenty-

five feet in length, and had long lashing tails and double-hinged jaws that enabled them to capture large prey, such as ammonites three feet in diameter, and other large cephalopods.

Our bathysphere is sitting in the depths of the warm shallow seas beyond the deltas. We are back in the Upper Cretaceous, 66 million years back in time. Ammonites are swimming among waving weeds, jettisoning water from their spiralled shells to keep moving. Some are heavily ornamented, with ridges and spines; some are plain spirals; some are spirals pulled out like corkscrews. Along with straight-shelled baculites with tentacled heads, they are feeding on minute marine organisms. Shoals of small fishes swim overhead. A solitary skate floats past in search of food; crabs scurry into the weeds; a carnivorous snail slowly moves along the sea floor.

An undulating shadow darkens the sea bottom. The agitated ammonites dart upwards and forwards. The large jaws of a diving mosasaur close in on one; crushing the shell, it swallows the flesh in a veil of ink.

The mosasaur, the ammonites and the marine organisms all have their days numbered. In the next 2 to 3 million years, all will become extinct.

Twelve years ago I wrote, "No trace has been found of cretaceous butterflies and wasps, cockroaches or grasshoppers, though these are known to have existed elsewhere . . . When there was so much sustenance, according to the fossilized leaves in the coal stratas, surely there must have been creatures to forage, insects to fertilize and termites to break down humus, so that more plants and trees could grow."

This information has changed. There were dinosaurs to browse on the leaves of the trees. There were insects. Petrified wood has recently been found containing termite galleries and excrement. And a concretion containing a cockroach wing has been found in Nanaimo. Indications have also appeared in the petrified wood of wood-boring ants. No trace, as yet, has been found of multituberca-lates—small rodent like creatures—that, as very early mammals, survived by staying small and hidden.

It is 65 million years ago. We are again above the moving map, watching the pebbles of the Gabriola conglomerate, interspersed with streams of fine sand, tumbling over the Spray Formation, now hardened into sandstone. Skies are darkening and the climate is colder. This new sediment is coming from low hills to the south of previous flows, and is creating what will be the St. John's Point Pensinsular, the extended foot of our Hornby moonshell. Further south, between Galiano Island and the mainland, sand is piling up, shutting off the sea from the south.

We are watching a period of transition. Sedimentation is now stop-ping, the seas are shrinking, the climate is changing and certain creatures are disappearing altogether. Toothed whales, sperm whales and porpoises are taking the place of mosasaurs and ichthyosaurs in the seas; birds and bats are replacing flying lizards in the skies; dinosaurs are bequeathing their world to small nocturnal mammals. The last of the ammonites is dropping its shell in the soft sand of the ocean floor.

Why is this happening? What is causing the death of the dinosaurs, along with the ammonites and other creatures?

Some propose that, during this time, an asteroid struck the earth, resulting in the skies darkening from the mass of dust stirred up, and the earth being covered with a layer of rare iridium.

However, the death of the ammonites, according to Cretaceous-Tertiary sedimentary strata in Spain, predates the layer of iridium also recorded in these rocks. And it took the dinosaurs perhaps 5 million years to become extinct.

An asteroid may certainly have struck, but other factors can have contributed to this change in the composition of the world's living plants and creatures.

Throughout the 70 million years of the Cretaceous, there was some plate movement, though generally the world was covered with shallow seas, low land masses and evenly distributed climate. At the end of the Cretaceous, perhaps due to a switch in the earth's polarity, the plates became extremely active, causing what is known as Deccan volcanism on the Indian Continent—not yet attached to Asia—and vigorous mountain building: the Andes, the Coast Range, and the beginning of the Rockies. The effect of so much volcanic steam and ash would have darkened the skies, lowered the temperatures and increased the rainfall.

Three quarters of the plankton in the seas died at this time. Perhaps the deaths were due to the deepening of the waters and shrinking of the seas from mountain building, perhaps due to ash from the volcanoes on the surface of the water or iridium from outer space. The disappearance of the lowest form of life in the food chain could well have set off a cycle of disaster, ending up with the largest animals losing their normal source of food.

The North American plate moving westward away from the European plate opened up the Atlantic and changed ocean currents. The creation of new mountain ranges affected wind currents. The temperature gradient, widening between the poles and the equator, brought increasingly cold winters and warm summers to this part of the world. Any of this could have contributed to a quantum change in the kind of life found on this planet.

About this time the seasonal migration of birds started.

Chapter Seven

63 TO 10 MILLION YEARS AGO

HELLIWELL PARK ON THE ST. JOHN'S POINT Peninsular has many moods. Visiting in late winter, we might find mist partly shrouding the grove of oak trees, and a grey stillness broken only by the trumpeting sounds of migrating murres.

Or we might find the south east wind, bearing rain or sleet, gusting up the Gulf, causing us to bend our heads as we battle the wind.

Dawdling in the evening light of summer, we might see clusters of stunted firs standing in the midst of rolling golden grass and have the feeling that the strange clear light is evocative of other epochs, other times.

But none of this magic is evident as we stand on the horizontal conglomerate rocks of the St. John's Point Peninsular, in the early Tertiary, 63 million years ago.

The skies are grey, the wind is cold, and there is the stench of death. In cold seas, sharks nose their way in search of food. Mouldering shells, green with algae, rest on the edge of the shrunken sea.

Layer on layer of Hornby's sedimentary rock stretches up to the stricken tree line, like a vast dead beach at low tide. Scavenging birds pick at pieces of decaying vegetation, abandoned tree trunks rest on dried-out rocks, while sluggish rivers seep through fetid pools above which insects swarm.

Tern-like birds still utter their cries over the water and, along the sea line, wading birds peck at snails and scuttering crabs. Suddenly, a shaft of sunlight pierces the clouds, adding a momentary smudge of colour to the greyness of the scene.

Millions of years pass. The skies have lightened. The climate is growing warmer. Throughout the world, mammals are evolving; some, such as lemurs, are working their way along the tree tops; others, resembling hippotamuses, are wallowing in swamps. Some, no larger than dogs, are browsing on ferns in the undergrowth. Closer to Hornby, we have a glimpse of cypresses, elms, evergreen trees, walnuts, holly and palm trees, growing where the Kitsilano area of Vancouver is now. There are oak, linden trees, maples and alders in the interior of British Columbia. The undergrowth is still not grass but ferns.

It is 53 million years ago. Looking down from high above Western Canada, we realize the large stretches of land surrounded by water that once stretched from Wrangellia to the North American plate have conglomerated into one enormous carpet of land. To the east, along what was once the North American continental plate's western shoreline, there is a seam of smoking mountainous rock, the Rockies, and where Wrangellia was joined to the Stikine Terrane, a smaller seam, the Coast Range.

As we watch, this carpet of land is being slowly rumpled. From the western shore line of Wrangellia over to the Rockies, land is being folded into hills. The sedimentary rock of Denman and Hornby is slowly being pushed up into two humps.

The Laramide Orogany, as these activities of the Pacific plate are known, comes to an end. New land, much of it one day to be known as British Columbia, has been buckled to the western edge of the North American plate.

It is 50 million years ago. The scene of tectonic action has switched from the Rockies to southern Vancouver Island. A new ridge, bubbling out magma from the earth's centre, situated off the West Coast of Vancouver Island, is creating a new plate, the Juan de Fuca. This plate is competing with the giant Pacific Ocean Plate, and carrying a new oceanic terrane towards the southern end of Vancouver Island.

We watch this newly arrived oceanic terrane being welded to the land to the south—today's southern Vancouver Island. The impact is tilting the sedimentary rock of the southern Gulf Islands, making the strata in some places rest at right angles to its original horizontal position. The pressure from the advancing plate is creating a series of ripples up the east coast of Vancouver Island. First northeast and then northwest, fractures and faults appear.

The folded hill of Hornby cracks and shudders, boulders of sandstone break away, conglomerate crumbles back into soil and pebbles get washed down towards the sea. Whole sections crack, and as they move sideways, drop to different levels. The bluffs, heights and benches of Hornby, the embryonic shape of today's Hornby Island, are gradually being created.

Where was the hammerstone when this was happening? Possibly still embedded in the Geoffrey conglomerate. Or perhaps, having been dislodged by an earthquake or by erosion, on its way down to the sea.

Today's newspapers warn that a big earthquake is likely to happen soon in this area of the Gulf of Georgia. The plate that is causing this tension is the same Juan de Fuca Plate that came into being 50 million years ago. Today it is being squeezed, together with an even newer plate, the Explorer, under Vancouver Island by the movement of the Pacific plate and the North American plate. This tectonic activity is also responsible for reactivating supposedly dormant volca-

noes, such as the Mount St. Helen's eruptions which took place in the State of Washington in 1980, and for the tilting of Vancouver Island towards the east. We read that the west coast of the island has risen as much as one foot since the beginning of the twentieth century. Hornby Island may not have been in the same latitude 30 to 50 million years ago, but the same rhythms from the same global pulse are still creating the same recurring patterns.

While we watch Hornby slowly changing shape, the average annual temperature of the world is dropping. In the Oligocene, about 35 million years ago, it was eighteen degrees celsius. During the Miocene, about 25 million years ago, temperatures dropped to fourteen to sixteen degrees celsius. We are told that this may be due to a tilt in the Earth's axis of rotation, as the Antarctic continent moves from a temperate zone into a polar zone.

The lowering of temperature is affecting the vegetation on the earth and the sources of food for the mammals. Woodlands are disappearing and grasslands increasing. With the increase of the grassland comes an increase in the mammal population. The creatures surviving this change in vegetation are long of leg and swift of hoof, ancestors of today's horses, deer and elephants.

It is 10 million years ago. There is no water in the Gulf of Georgia. All we can see as we stand on the top of Mount Geoffrey, and look across to Texada, are rolling grasslands and clumps of deciduous trees. Directly below us to the west, a family of ambledons, shovel-tusked mastodons, are picking their way across the River Lambert. Galloping over towards Lasqueti are a herd of merycodus, now-extinct pronghorn antelope. Close by, perhaps where Olsen's farm now is, are groups of grazing pliohippes, the ancestors of today's horses.

A friend who found a tree of great antiquity on the island asked the tree if she might speak with it. I was with her. She leaned against its rugged trunk and closed her eyes. After a moment or two of silence, she spoke of what she saw: rolling grasslands with clumps of trees, across which were grazing herds of strange animals, some as large as elephants.

She did not know that several million years ago the Gulf of Georgia was dry. I later showed her a drawing in a book of pre-history which described this period. "Yes," she said, "that is exactly what I saw."

The skin and padding of the island keeps changing; the soil that grew the grass on which these creatures grazed will have been blown or washed elsewhere; the contours of the Geoffrey Hill will have altered several times from natural erosion and from the comings and goings of the glaciers. But the rock that is Hornby Island today—rock that we walk on, sketch, wonder about—was here through all these changes.

If all is recycled and nothing destroyed, then the hardpan of Hornby, in which the ancient tree had its roots, may have been here, perhaps as rock, when Hornby was a hill in the midst of grasslands. There may be a line of communication, which my friend picked up, that we do not at present understand.

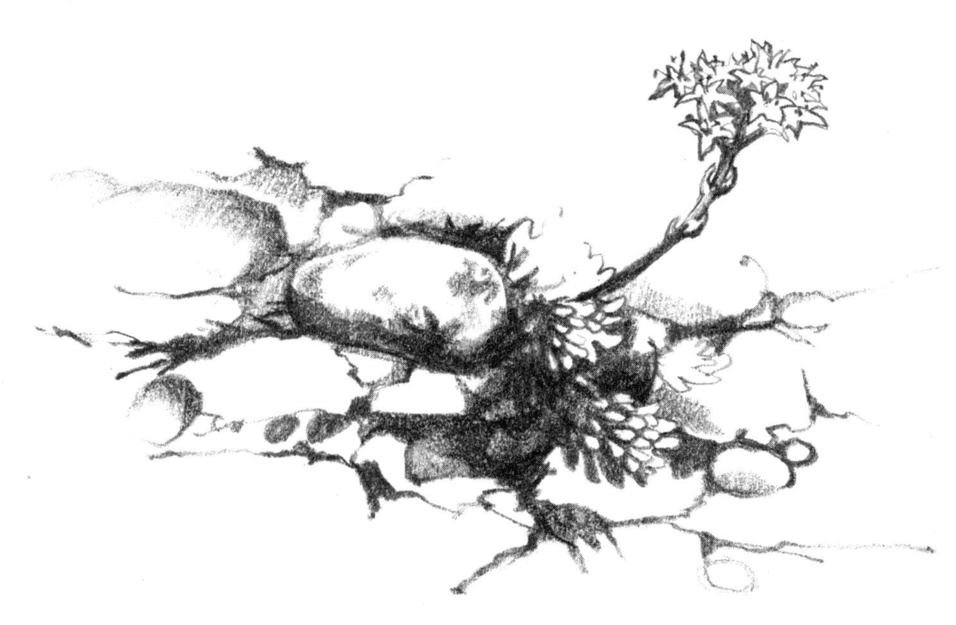

Stonecrop, the little yellow flower that brightens Helliwell in spring.

Chapter Eight

2 MILLION TO 9000 YEARS AGO

HORNBY IS ABOUT TO EXPERIENCE FOUR ICE AGES and four interglacials. The fourth interglacial is from 10,000 years ago to where we are now.

What will come after "where we are now"? Will there be, as we once thought, a fifth ice age? How is global warming, caused by holes in the ozone layer, going to affect future vegetation? And if the seas rise from the melting of the ice caps, what will the future contours of Hornby be like?

We are 2 million years back in time. The climate is already cooler than it was in our last glimpse of Hornby, and the seas have returned. Autumn leaves are being frosted before they are ready to fall. Early spring buds are nipped by extended cold. Warm seasons are shortening. Animals are searching for new kinds of fodder as their familiar diet slowly withers away. Some adapt, some die, some move towards the south.

As the conifers grow thinner and sparser, the rivers Lambert and Baynes freeze, first only in winter, then all the year round. The land stays frozen, holding the snow which is falling now instead of rain. The frozen earth freezes the roots of trees. Eventually, only plants of the tundra are surviving.

Over the frozen, dying land, the glaciers creep. First from Vancouver Island, then from the mainland mountains, over the top of Texada, the glaciers keep adding to their load of ice. Dirt, gravel, pebbles scoured from hill tops, and boulders broken and cracked from the land, join the slow procession.

The glaciers come and go several times. During glaciation the weight of ice depresses the island and all surrounding land. The land rebounds in the interglacials, but, with the increase of water from the melting ice, during much of the interglacials, Hornby is under water.

There is a sequence of events and a repeating order to the movement of the glaciers and to the pattern of deposition of eroded materials—first ice debris, then sea debris, then changing tide levels—that is reminiscent of the laying down of the cretaceous sedimentary rock. The time span is different, the agency for erosion is different, but the same relatively orderly pattern of renewal persists.

It is about 65,000 years ago. The Geoffrey rocks are several thousand feet beneath a thick blanket of ice. It is summer time at the height of one of the glacial advances. Snow fields stretch for miles in every direction. The whiteness is broken towards Vancouver Island and also over towards the Coast mountains by the black tops of bare mountains. The weather is clear, the sky a brilliant blue. The summer sun reflects off the snow. It is very hot. From midday on there is a barrage of sound from distant avalanches thundering over glaciers. Towards nightfall, the sun, with a flush of green light, goes down over Quadra Island. Cold moves in. Within minutes, a sheet of ice is everywhere. Icy winds blow.

It gets warmer. In the south, land and sea start to reappear. The glaciers as they retreat are dropping their burdens of boulders, gravel, silt, clay and sand, filling in depressions, covering scoured areas.

At the northern end of the Gulf of Georgia rivers coming down the inlets of the mainland are in full spate, washing out into the Gulf tons and tons of white sand. A thin skin of ice covers these waters. Under the ice and above the salty seawater, the river water travels carrying quantities of sand. This potential soil is blanketing the eastern shores of Vancouver Island, from Quadra Island south.

What will one day be the Willemar Bluffs at Comox, and the Komass Bluffs on Denman, are being created. On these massive deposits of sandy soil, vegetation starts to grow. From Johnstone Strait down to the southern parts of the Gulf of Georgia, the chequered scene is one of melting glaciers, flooding waters, and stretches of new land, green with burgeoning plant life.

To the northeast, over the grey tops of the mountains swathed with wisps of cloud, we look to the interior plateau and on to beyond the Rockies. On the prairies, which do not look too different from today, if we watch closely, we can glimpse coyote, bison, skunk, rabbit, horse, camel, giant ground sloths and mammoth. For the moment we see no sign of people. But they could be there.

At the start of the ice ages, nearly 2 million years ago, homo erectus spread from Africa into Asia and Europe. About one and a half million years ago humans discovered fire. This enabled them to live in colder climates, feed themselves better, and be a little more protected from predatory animals and the terrors of night. About 60,000 to 50,000 years ago (the time at which we are now looking), homo sapiens are said to have reached Australia. During the ice ages there were at least four periods when hunters could have crossed the Bering Straits.

We are back above Hornby. It is 40,000 years ago. Sea levels have dropped. The layer of thin ice on the sea has disappeared, the climate is warmer. Now, and for the next five thousand years, the last interglacial is at its warmest. The trees and undergrowth on Hornby are similar to what we see today, though the sea level is higher.

An analysis of a landslide at Dashwood, just north of Qualicum, close to Hornby Island, has enabled scientists, with the aid of carbon dating and the taking of spore samples, to reconstruct the climate, vegetation and the sea levels experienced here during this last interglacial. They found evidence of pines, Douglas firs, birches and poplars, grass, ferns, and plants of the daisy family.

A complete walrus skeleton was found near this site several years ago by Graham Beard. It was restored, mounted, and dated as 70,000 years old. The skeleton is on loan from the National Museum of Natural Sciences in Ottawa, and can be seen in the Qualicum Museum.

The climate grows colder, and the trees thin out. As they slowly die from long winters and cold summers, dead trees tumble into Strachan Valley. (Can these be the same trees that James Strachan kept finding embedded in the mud, when at the beginning of the twentieth century, to create farmland, he drained the lake in the valley?) Grass, sage brush and marsh plants become more abundant; then they too disappear under a permanent blanket of snow.

Glaciers, travelling southeast from Vancouver Island, are once again scouring the Geoffrey conglomerate. Newly arrived glaciers, travelling down and across the gulf from the mainland are carrying rocks and pebbles, sand and boulders. A large granite boulder travels from the Coast Range across the white wastes of the gulf. As it moves, it accumulates a small pimply, "golf-tee" support of ice and snow. The boulder topples. Still moving, a new "golf-tee" support is made. Large granite boulders are familiar land marks on the island today. Under the moving ice there are cold seas in which fish and other deep water creatures swim.

Back in the house, above the cabin, we look at some tiny fossils of the swimming scallops and other deep sea creatures, brought to the surface by a well digger from the Raven Lumber subdivision, close to the cabin. They were found in clay under hardpan. This means that 20,000 years ago, the land where the cabin now stands was covered with several hundred feet of icy water.

People have not reached Hornby yet, but somewhere to the north, small dark figures are moving south. They are crossing cracking ice-floes, battling snow storms, fishing through the ice, clubbing and spearing sea-mammals. To the southeast, down the Columbia River, there are other moving figures: people following the animals migrating from the prairies in search of pasture, using stone tools, perhaps less delicate than those used by their cousins to the north; people hunting deer, trapping birds, collecting mollusks, and later, as they reach the sea, clubbing seal and catching salmon.

It is 15,000 to 13,000 years ago. The climate is changing. The glaciers are again retreating. The inside of a glacier moves faster than the outside; glacial fingers leave banks of gravel.

The island, without pressure from the weight of ice, is rising, scarred and barren. On the island's lower flanks the subsiding sea, thinly covered with ice and thick with sand, clay, mud and pebbles, is dropping its burdens, filling in ravines and crevices, spreading a kind of mud and plaster everywhere. At just about 150 feet above our present day shoreline—about 11,500 years ago—the retreat of the sea stops.

Today, evidence of a beach on the 150 foot contour line dating from this time has been found. Large clamshells, buried nine feet under sand, have been dug up.

From the melting of a million glaciers the sea begins to rise, and the island is again submerged. This time the rising water is not saturated with glacial till, but with sand and clay from flooding rivers. The water rises and rises.

It is 10,500 years ago. We are again on the conglomerate rock of Mount Geoffrey.

Hornby is now a small boomerang-shaped island, no more than a mile in length and a quarter of a mile across; its highest and most narrow point is in the centre. It is barren except for tufts of grass, moss and lichens. For miles in every direction there is icy water.

On the cliff, facing the small hills of Vancouver Island, sea birds are nesting in crevices. There is no sign of Denman Island. Looking towards the mainland, we see Texada Island as a string of islets. Only the tip of Lasqueti shows. The journeyings of the glaciers have left the rock surface scarred, with scatterings of boulders and rocks from both Vancouver Island and the mountains of the Coast Range. In the pockets of newly deposited till, small plants are starting to grow.

For the next 1,500 years—10,500 to 9,000 B.P.—the sea is subsiding, and the island growing bigger. The movement of the retreating sea is packing sand and clay into low areas, covering gravel and levelling out depressions.

Traces of another ancient beach, dating to about 10,000 years ago, have been found on the northern shore near Tralee Point. Here, about fifty feet above sea level and 300 feet inland, the digging of a shallow well brought up, from about three feet down, large clam shells, as fresh and clean as any we dig today.

Myth has it that Hornby Island "rose from the sea." It did, several times. The last time was 11,000 to 9,000 years ago. How did the tellers of myth know?

PART TWO

First People

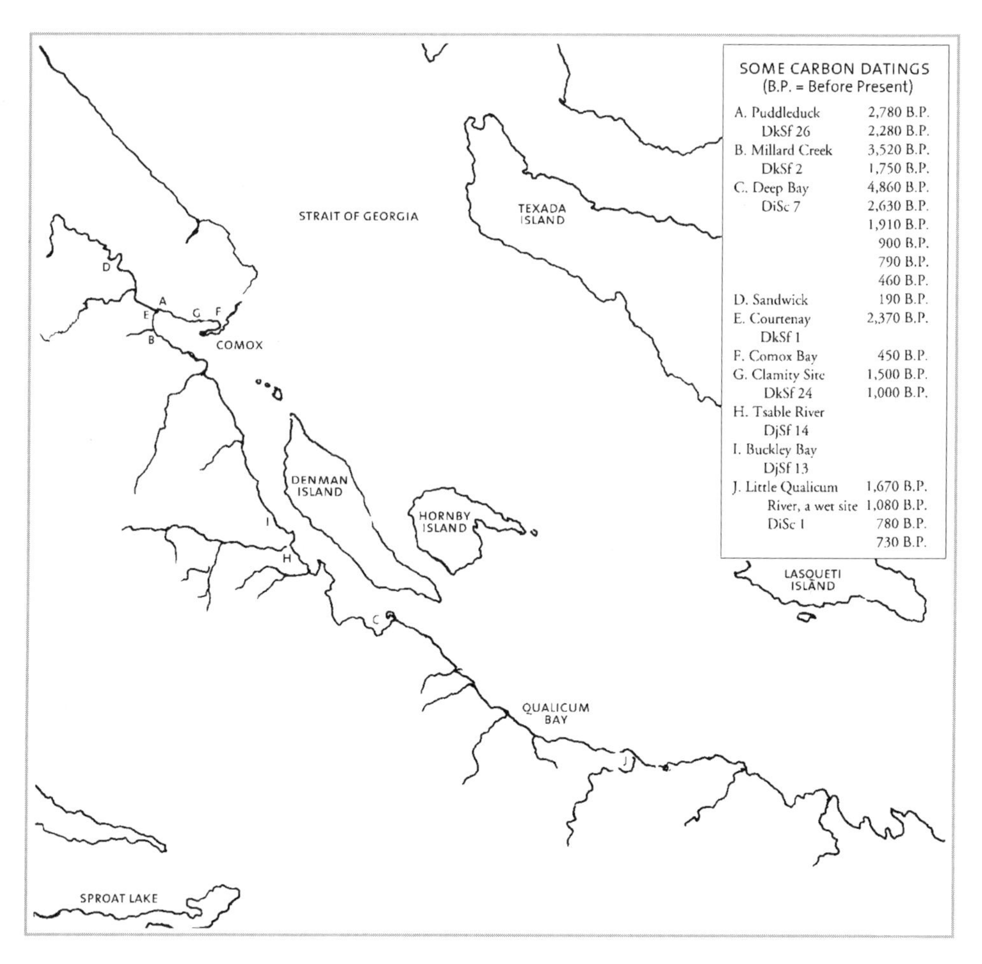

Carbon dating from archaelogical digs in the vicinity of Hornby Island.

Chapter Nine

9,000 TO 1,500 YEARS AGO

NINE THOUSAND YEARS AGO. THE OUTLINE OF HORNBY Island is almost recognizable as the island we know today—a reclining lion, if seen from Vancouver Island; a moon snail with foot extended, if seen from the air. The beaches are smoother, there are no sandstone sculptures, the contours are rounder. First pine trees, now alders are becoming established.

While Hornby is in our immediate focus, our eyes can see as far north as Bella Coola, and as far south as the Fraser Valley. What we are watching is no longer the dramatic geological movement of land or of ice, but the gentler, slower development of an ecosystem, one that embraces people.

Archaeology is still a young science in British Columbia. Hundreds of thousands of cubic feet of prehistory lie along our shores, waiting to be deciphered. Much has been lost forever to developers. Excavations have been made in Courtenay, Comox, Millard Creek (near Royston), Buckley Bay, Tsable River, and Deep Bay. Hornby was in the territory of these people, so that information which has been discovered from these digs can legitimately be applied to Hornby.

There have been no official archaeological digs on Hornby, but there are about eight feet of midden at the Shingle Spit and about four feet at Ford's Cove, and many pockets and stretches of midden both along and above the shoreline. Artifacts have been found on the island: bowls and pestles at the Shingle Spit, knives and sharpened stone at Heron Rocks, pestles at Sandpiper, the hammerstone near Tralee Point, arrowheads everywhere. But out of historical context they tell little.

Far to the north at Namu, near Bella Coola, hunting and fishing people have made a settlement. To the southeast, people coming down the Fraser Canyon have met the sea covering what is now the Fraser Valley, and can be seen chipping stone. As the sea retreats down the Fraser Valley, the people follow.

It is 8,000 years ago, there is a settlement at Bear Creek, Northern Vancouver Island. We do not know if the people arrived from the south or from the north.

As seasons merge into seasons, and a hundred years are followed by a hundred more, we watch the rainfall increasing, the water courses becoming small streams. Douglas fir is added to the pine and alder, trees and undergrowth grow bigger and more lush. The sea continues to subside.

The contours of Hornby Island enlarge, and what is today an intertidal zone is covered with soil, grass and trees.

On this new land slowly reclaimed from the sea the occasional group of people camp. They drink spring water, gather immediate needs from forest and shore, scan the surrounding scene from the top of Mount Geoffrey.

Because the sea level, 7,000 to 5,000 years ago, is lower, no trace will now be found of these habitations.

It is always along the shore that the first people live, for the sea is all important. The sea provides fish and sea mammals for food and bone tools.

The sea reveals, between tides, cobbles for shaping and sharpening; shellfish and seaweed for steaming and eating. The sea is the highway and the link with other families, with other gathering areas, and with the larger world beyond. The heavily forested land behind the shoreline is fraught with unknown dangers, real and mythical. No one lives there.

It is 6,600 years ago. Far to the south, Mount Mazama in Oregon is erupting. Intermittently, for hundreds of years, dark clouds containing steam and ash move north on the southeast winds. Rain keeps coming down. It is cold. The rocks of Hornby shudder from a succession of earthquakes; cliffs tumble, sandstone slabs crack. In the Fraser Valley ash is built up over the land. This layer of ash today helps date archaeological finds.

It is 5,000 years ago. The air is clear, the climate warm. The rain has almost stopped. The sea level has risen. The land of Hornby is park-like. On its southern slopes there are larger oak groves, more clumps of arbutus, more grass and cactus than today. Over most of the island the forests are of fir, hemlock and spruce. There is no cedar.

There are people at Deep Bay, across the water from Hornby, on Vancouver Island. They are chipping at stone tools, bartering with traders for obsidian, a translucent volcanic rock, not steel but the next thing to it. The obsidian probably came from Anahim Peak, near Bella Coola, but it could have come from as far away as Mount Edzia in northern British Columbia, or from Oregon in the United States.

They paddle over to Hornby Island to dig for blue camas bulbs in the summer and to collect acorns in the fall. Some are fishing off Flora Island and Norris Rock. Back in slowed-down geological time we watch the sea continue to rise.

Today in a midden at Deep Bay, dating back to this time, (a time known by archaeologists as the "Lithic Period"), there is evidence of stone tools. If shell and/or bone was also being used for tools, this evi-

dence has not survived. Relics may have been washed away as the sea level continued to rise.

We are now in the period known by archeologists as Locarno. Sea levels are fifteen feet higher than today. On the Comox Estuary, halfway between present-day Courtenay and Royston, women on wet, pebbly shores are collecting mussels. Distantly, we hear the cry of men mingling with the harsh call of gulls as fish, wriggling and glistening, are drawn up in nets.

There is more rain. The delicate tracery of young cedar starts to grow among the fern and bracken in the forests. Streams are noisier. Water lies in swamps. New land is being slowly built up around the shoreline, as the sea level again retreats.

Mists evaporate into sunshine. There are settlements in the Comox Valley, at Millard Creek and at Deep Bay. Women are sitting in groups above the tideline, with the dark forests stretching behind, weaving mats and baskets with bone awls, stitching hides and skins with needles made from mammal and bird bones, cooking in deep earth ovens. Near their drawn-up canoes men are squatting: one is repairing a net, another binding a bone point into a fish hook. A family takes canoes from Millard Creek to Tree Island to gather clams.

The cedar trees have now grown to maturity. People are stripping bark, digging roots, splitting pieces of trunk.

On the north coast of Hornby to the west of Tralee Point there is water and shelter from the winds. As a south east storm sweeps up the gulf, some fishermen bucket ashore and beach their canoes.

On the south side of the island, above the Shingle Spit, an occasional man, sometimes young, sometimes old, comes ashore, seeking solitude. The younger men may be on their spirit quests. The older men may be shamans looking for purity in the island forests.

Parties of fishermen or hunters, and canoe loads of gatherers and root diggers come to the island. At first they appear to come haphazardly and for short periods. Later, their numbers grow. Campsites are used for longer and longer periods. A pattern seems to be emerging.

It is now 2,500 years ago. To the south, at Marpole on the mouth of the Fraser River, there is a large settlement.

People here are building houses with large house posts and fashioning massive canoes (the cedar forests have now reached maturity). People are coming and going, strangers are being welcomed; trading is taking place. Resources and decorated goods are being given in return for an apparent excess of salmon. The water is ruffled as tens of thousands of salmon surge in the fall of each year up the Fraser River to spawn.

People are coming in canoe loads from the Gulf Islands and from the State of Washington, and on foot down the Fraser Canyon from the interior, trading dentalia from the west coast, obsidian from the south, jade and soapstone from the interior.

Also down the trade routes have come ideas. The smoking of tobacco in pipes, with the trading of decorated pipe bowls. Labrets, lip ornaments used here since Locarno times, are also a phenomena to be found among primitive people in Africa and South America. We are now moving into the Marpole period.

For about a thousand years, 500 B.C. to 500 A.D., looking south from Hornby, there appears to be an affluent society near the mouth of the Fraser. It has a class structure, a use for non-functional goods.

From a distance we watch Marpole people weaving baskets, steaming bent wooden boxes into shape, decorating wooden bowls and canoe paddles, fashioning ornaments and sculptures from soapstone and coal, giving deference to their wealthy leaders, snubbing their enslaved captives. Some of their ornaments, weapons and exotic materials, such as copper and amber, are found today in the elaborate burials accorded to high-ranking people.

It is now 1,500 years ago, 500 A.D. Marpole as a thriving centre is fading away. There is much less activity around the mouth of the Fraser.

Instead, we see settlements on the southern Gulf Islands, and up the coast to Comox.

Anthropologists will tell us that a simpler, more egalitarian, culture is evolving, a culture called by them the "Gulf of Georgia." Within this new culture, there are still traces of the old. From Quadra Island in the north to Washington State in the south, we are beginning to see the shape of a loosely related network of people. A network which will be called, 1,450 years later (in our concept of time), the "Coast Salish."

It is 1,100 years ago, 900 A.D. Across the water from Hornby on Vancouver Island, at Deep Bay, some of the ancient Marpole culture is being preserved. A child is being buried in a cedar box, together with a large jade celt (an axe which would be hafted into a wooden handle), a pendant shaped like an animal, two copper beads, dentalium and shell beads. This is a Marpole burial pattern. Who is this child to be afforded such honours?

Her skeleton when found was incomplete. How did she die? And when alive, did she visit Hornby, run on the sands of Tribune Bay, or know what it was like to swim in the deep pools of Downes Point?

Chapter Ten

500 TO 1790 A.D.

WE ARE BACK ON THE ISLAND IN THE TWENTY-first century, watching layer on layer of a midden being uncovered. Measurements are being taken and notes written about each depth. All the soil is being screened.

We have found an awl—an implement for making mats—and at different depths several ancient fireplaces, circles of blackened stones. Piles of clam shells, neatly stacked together like discarded egg shells, keep appearing.

One ring of stones has a beautiful white collar of crushed sea urchin shells surrounding it. Who was the woman that did this? Was everything undertaken by the early Hornby people done with such caring?

We are back on top of the mountain in seasonal and cyclical time. It is early spring any year between 500 A.D. and 1790 A.D. (our time).

Far below us is the sand and gravel of the bay at the Shingle Spit. Up and beyond, divided by the dark shape of Denman Island, are the silver waters of the Lambert Channel and Baynes Sound. Light is shafting through veils of white cloud, giving brief glimpses of the Beaufort Range.

All sense of linear measurement is fading. We are aware of the rat-a-tat-tat of a nearby woodpecker, of moving shadows on the trunk of a tree, of a slight wind blowing from the north, but all the twentieth century boxes of knowledge and ladders of achievement from which we get our usual sense of security have, for the moment, gone. Days, seasons, tides, are spiraling around us, in renewal patterns of birth, death and rebirth. The circles are moving at different speeds as if to form a ball.

From far away from across the water comes the sound of dipping paddles, and of excited voices calling. Rounding the south end of Denman, below and to our immediate left, is a flotilla of canoes—small canoes with one person paddling, larger canoes with three or four people in them, twin canoes, joined together by cedar planks.

It is spring time, and the Pentlatch people, the Coast Salish people who have spent the winter in their big houses on Vancouver Island, are paddling to Hornby. For the next nine months they will camp on the island, gathering, hunting, digging, preserving, fishing.

We land on the pebbly beach of the Shingle Spit. The fresh spring air is full of the sound of splashing, bumping, hammering, calling, and laughing. From the canoes resting on the shore, men are carrying split cedar planks and are creating a house by tying the planks with twine to poles already in position, poles left there from last summer. Women and children are running back and forth with tools, stores, bedding, matting; smaller children are collecting chips of driftwood from the wrack of seaweed left by the winter tides.

We move into the trees and . . .

. . . back into the twenty-first century. The grass beneath our feet is mown, and there is a shuttered white cottage behind us. On the left, sunshine is warming the blue roof of the pub and restaurant. Beyond a massive limestone breakwater, we see the ferry coming in. Its

engine noise changes, the motors go into reverse. With a splash and a bump, the ship touches the black piling, and the ferry noses into shore. With a loud creaking, the ramp is lowered.

Helmeted motorcyclists zoom up the slope, forerunners rev their engines . . . station wagons with plastic totes roped on to their roofs, trucks crammed with building materials, campers, and polished city cars, are moving on to the island, along the granite breakwater. We sense excitement, a feeling of arrival, and of escape . . .

. . . excitement and escape! The people are glad that the long winter, confined in the houses of the winter villages, is over. Left behind are the long dark days with perpetual rain, the tensions of living in close quarters with relations and in-laws, the growing lack of fresh food, the stormy weather that prevents journeying anywhere, the slippery beaches. Ahead are several months of gathering, hunting, fishing, preserving, shifting camp, and freedom!

Tonight, instead of sleeping on raised platforms around sunken earth floors under a gabled roof, each family section partitioned by matting from the next, with the air heavy with the stench of the past winter, the people will be sleeping on matting placed over soft ferns, with the spring breezes filtering through the crevices in the shelter, carrying the scent of the sea.

Evening comes. From several small fires wafts the smell of wood smoke and fish cooking. Streams full of spring rain are gurgling down the side of the mountain. From across the water, a sea lion calls.

In the twentieth century, the sound in spring of a sea lion calling was something to stop and be excited about. Now in the twenty-first century, sea lions seem to be with us year round. In the spring, the rhythmic sound of lions barking resounds right across the island,

and some islanders enjoy taking off in small boats from Ford's Cove to watch the sea lions swimming, diving, lolling and pulling themselves around on the rocky islets off Hornby' s southern coast. Why have the number of sea lions increased so much in the last few years?

It could be the diminishment in the numbers of their predators, the orca. It could be shortage of food in the north due to heavy pollack fishing off Alaska; it could be warming of the waters. The age-long pattern of sea lion migration seems to have changed. Not only are there more, they stay in these waters much longer.

Physically the island is familiar. It is the relationship of the people to the island that makes everything different. We only sometimes tune into the unseen world. Pentlach people are a part of it.

The familiar world for them starts at Comox and, taking in Denman and Hornby Islands, stretches along the coast to Qualicum Bay. In all directions beyond this there are dangers, fierce spirits and warlike people.

Boulders, headlands, special trees, the creatures on the beaches and in the woods, the stars overhead, all have special myths. Myths that instruct, warn, explain, and give a sense of continuity.

In seasonal, cyclical time, the island is nearly all forest, with trees much taller and more widely spaced than today, and carpeted with ferns and mosses. There are areas of swamp, grassy glades, streams that run year round. There are more fish in the sea, more clams on the beaches.

Almost every bird call, rustle of the wind or change in the sound of the sea, has a message. Almost every tree, bush, plant, moss, seaweed, rock or pebble has a meaning and a use. And encompassing everything there is for them the presence of "spirit." Awareness of the mystical world permeates their lives as totally as the air they breathe.

It rains in the night. The people wake early. We see some of them swimming in the cool water of the bay, some chewing on dried deer meat, some taking off early to fish. Smoke spirals up from fires.

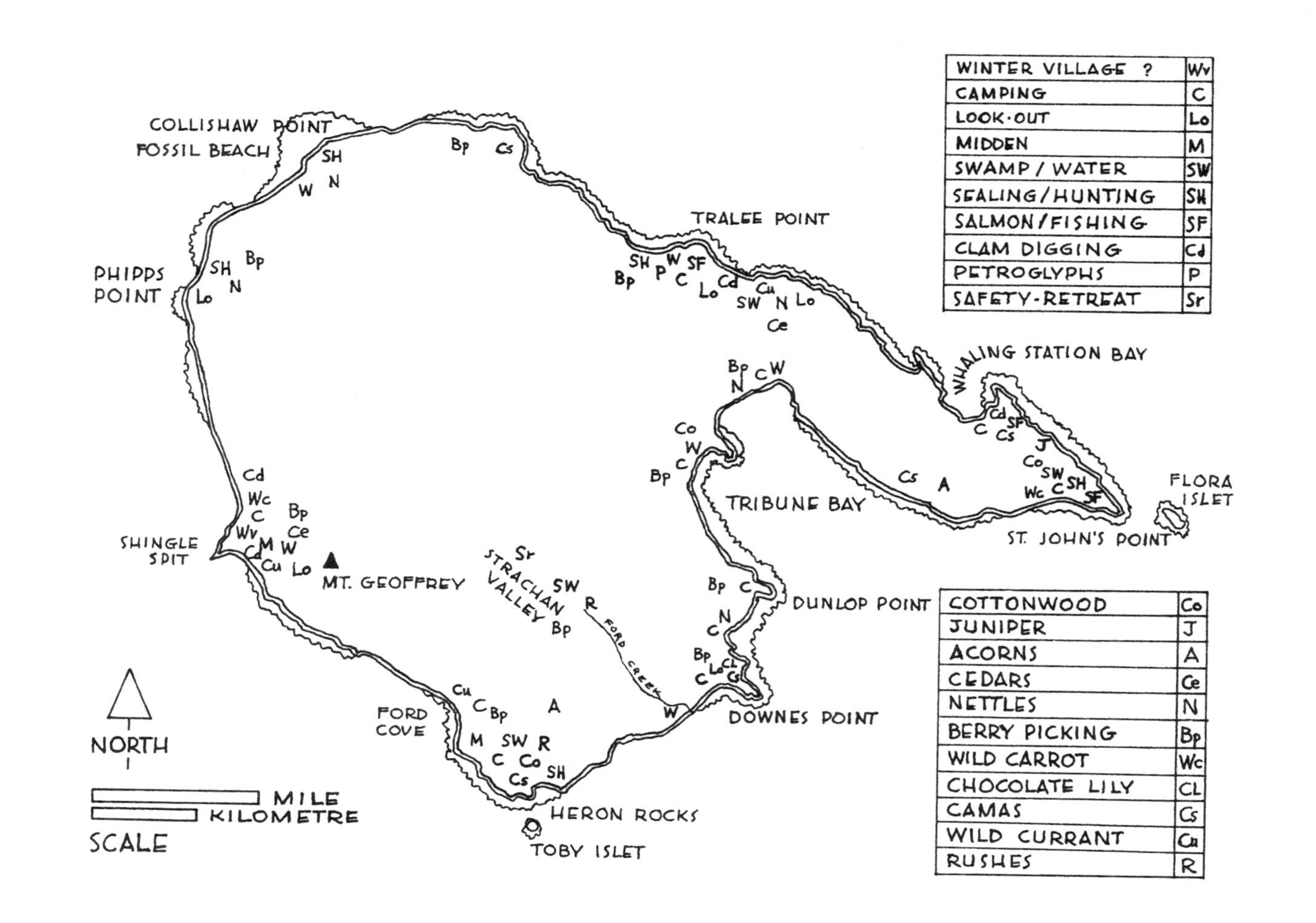

Hornby Island: An important part of the lives of the First People.

Chapter Eleven

EARLY SPRING

SEAWEED TRACES DELICATE PATTERNS ON THE *sands of Tribune Bay. It is early spring in the twenty-first century. Winter storms have brought in fresh logs and swept the beaches clear.*

We walk to where large concretions, six feet in length, lie embedded in sandstone. What decaying organic matter created them? Were they large creatures 66 million years ago? Or massive tree trunks?

Behind us, beyond a broad raft of driftwood logs, is a large area of rough grass. It is an area cleared for farming by early pioneers, now part of a provincial park. Encompassing the large field on three sides are woods, dark green cedars and firs, lightened by the spring greens and yellows of burgeoning deciduous trees, shrubs, bushes.

There is the sound of splashing, followed by the rasping sound of something being dragged over sand. Two young women are pulling a canoe onto the shore. Their long black hair is braided in two plaits. They wear over their shoulders elbow-length rain capes of sewn tule, and skirts of the shredded inner bark of cedar. They take baskets out of the canoe.

There are only a few driftwood logs on the beach and beyond, instead of a large field, there is a burnt-over forest. A few years earlier than this moment in time, a fire behind both Tribune Bays destroyed the usual forest, and a new cycle of ecological renewal was started. Amid the dead, charred trees the preliminary cycle of thistles, nettles, fireweed and horsetails has been followed by salmonberry, thimbleberry and rose bushes. Towards these bushes, now softly greening the foreshore, the women are heading.

They pick the young shoots of the salmonberry. They may also gather anything else they come across, such as the rhizomes of licorice fern, which after the long winter are to them what the first asparagus is to us. Or they may dig for the roots of silver weed. Elsewhere on the island women are digging wild carrots, careful to take only the largest roots and to replace the turf. None of them will pick the dark green rosettes of the nettle, as we would. These are left to be harvested later when the stalks can be twisted into fibre for rope.

The two women are walking back to their canoe, their baskets full, when one of them stops and picks up a large brown pebble. She examines it happily, then slips it into her basket. Later she will pressure flake it into a scraper.

The people who come to this island are semi-nomadic. For nine months of the year they use the island as a base from which to fish and as a living storehouse from which to gather almost all their needs.

Behind them are three to four months of living, with three or four other related families, in the big house in the winter village, away from storms and marauding people, away from the island. Ahead are many months of camping, and moving camp, of fishing, hunting, gathering, drying, collecting, processing, and feasting. Waiting for rain to stop, for wind to drop, for visiting and for being visited. They will also periodically go back to the winter village to store some of their harvest or, if there is a spell of bad weather, to take refuge.

Within the apparent flow of life, there is a complicated system of social and economic checks and balances. There are rituals that must be followed, unseen spirits to please.

Something should be happening, back in the twenty-first century. But where are the herring? Twelve years ago at this time of year, boats of all kinds, trollers, seiners, fishpackers, would have been gathering, anxiously waiting for the season to open. There would have been throughout the day the sound of small planes, trying to detect from the turquoise blue patches of herring spawn where the opening would be. On all the tall trees eagles would have been waiting, and on the rocks thousands of seagulls. But now on this beach all is quiet, a ruffle of water and a swirl of seagulls perhaps further out in the Gulf. A large seiner searches the shoreline. From its sounder, it is getting a clear picture of the marine activity below. We hear there has been an opening for an hour or two in the Lambert Channel. Over the last few years, all the Georgia Strait herring have disappeared. What is now being fished are herring that have travelled hundreds of miles from the Pacific Ocean in order to spawn.

At first it is only a small patch of turquoise water here, followed by a ruffle of water there. The eagles reconnoitre, and gulls screech. Men and women shout news to each other from passing canoes. The areas of turquoise water grow larger.

Lambert Channel is seething with silvery fish, and milky with spawn. The herring are here! Wings whirr, as birds thrust, snatch and chase, paddles dip, herring rakes splash, excited voices call.

A woman paddles, the man rakes. Flip, a wriggling fish, empaled on the rake, drops into the canoe. Swish, back into the water dips the rake; flip, more fish are tipped into the canoe. Seals and sea lions plunge and dive. White foam is washed up on the shore.

Along the shale and mudstone beach, between the Shingle Spit and Phipps Point, and along the rocky conglomerate shore between the Shingle Spit and Ford's Cove, and again across the Channel on Denman Island, bare-footed, bare-chested, women and children are breaking off branches of cedar and hemlock and laying them along the tide line to catch the translucent eggs of the herring spawn as it comes in on the tide.

Calling shrilly, mink are chasing down the beaches and into the water, scampering back into hiding with fish in their mouths. Slower moving racoon sit by the water's edge, solemnly and two-handedly stuffing themselves.

The tide retreats. People are gathering the fronds.

At Ford's Cove and the Shingle Spit and again at Heron Rocks, canoe loads of still-wriggling fish are being emptied on to shore. Emptied, filled, and then emptied again. For the next few days those people not out in the canoe are on the beaches, splitting the herring open with sharp stone knives, shooing away thieving ravens, turkey vultures, eagles, crows, gulls, hanging the split herring on racks to dry.

A few days later herring is sighted on the northern shore. Canoe loads of men, women and children paddle around to Tralee Point. Temporary camps are set up. This time circular stone walls will trap the herring. The tide comes in thick with dancing fish. Outside the traps, men in canoes direct the herring towards the shore. Inside the traps, knee-deep in water, women stand, thrashing with cedar branches, stopping fish from escaping. The sea goes out, the stranded herring and the spawn are gathered. The processing of the fish continues.

The taking of the herring after permission to fish has been given is fast and efficient. Most of the smaller boats have disappeared; most of the seiners now belong to one company.

Herring are being caught not to feed people on Hornby or even in British Columbia. They are caught for the extraction of the roe for export to Japan, where it is considered a great delicacy. Very large sums of money can be made. The flesh of the herring is considered secondary, it is turned into fertilizer. The main competitors with the fishermen for the herring are the sea lions, seals, eagles, seagulls, seabirds, which for hundreds of years have depended on the coming of the herring to give their diet the necessary boost before the arrival of their young in the spring.

Back in the rhythm of cyclical time, the spawning of the herring means more than feasting and excitement, and more than having dried herring for times of need. The herring do not spawn in profusion everywhere.

Some of the people hauling in herring will have come from other gathering territories. For this courtesy they will now be indebted to those of the Pentlatch people who "own" the water around Hornby. The Pentlatch people will also give surplus fish and spawn to kinsmen living outside the spawning area. These actions, besides ensuring there is no waste, will strengthen inter-family ties and build up credit for the future.

We join the people on the beach as they scrunch herring spawn. Daytime noises dwindle. There is the sound of crackling fires, of laughter and of voices murmuring. Then the night takes over. From a nearby swamp, the croaking of a thousand frogs.

Women in the forest digging cedar roots and coiling cedar bark.

Chapter Twelve

LATE SPRING

THE DAYS ARE GROWING LONGER. IN THE EARLY *morning, the warmth of yesterday's sun still lingers, and there is a smell of the sea.*

Around swamps and where the forest has been thinned there are smudges of white cherry blossom. Greens of every shade blush over the willow, alder and maple trees. In clearings deer are grazing on the new growth of soft green grass. Humming birds zip backwards and forwards amongst the small pink flowers on the green salmonberry bushes. After rain the air sparkles and shadows move.

Out from the shore surf scoters and scaups form and reform in their hundreds, making patterns on the smooth surface of the sea. There is a ruffle of water as a whole section dives. A jet plane screams across the sky . . .

. . . back in cyclical time, sunlight, and sudden showers follow dark days with threatening storm. Temporary camps are set up and taken down. Sometimes, it seems all the men are together and all the women are somewhere else. At other times, several small families are living in quite separate places, chiselling stone, mending nets, building fires.

A man, bare-chested, head banded with cedar bark, wearing shredded bark on a belt around his waist, beaches his canoe on the mudstone shore of Fossil Beach. Four women, baskets on their backs, walk in the forest sharing the paths of the deer, picking swamp grasses, searching for herbs, collecting, as they chance on it, wood for carving, rocks for fashioning, bark for binding or dyeing.

Type and depth of soil, water courses, prevailing winds, as well as sunshine and rain, all contribute to what is growing where and when on the island. While the island is yielding bounty all the time, what can be harvested varies from place to place as well as from season to season.

On the wide stretch of land to the west of the Shingle Spit, trees have been nourished for generations by water coming down the Geoffrey Escarpment and into the soil above the interglacial compacted sand covering the underlying Northumberland mud-stone rock. Now that it is early summer, and the sap is rising in the trees, people from the encampment at the Shingle Spit harvest bark and dig roots in this forest. They also hunt deer. The damp soil provides pockets of rich pasture.

When this land was cleared by pioneer farmers at the turn of the century, they found many arrowheads. "Relics of a battle," they thought. But the arrowheads told not of battles—bows and arrows were not used for this—but of generations of hunting. There is no evidence of the Coast Salish people being warlike. Unlike the tribes to the north, they had no warrior class. Only in the nineteenth century were fortifications needed and built.

Into the quiet of the forest, to the west of the Shingle Spit, we walk unseen behind two women. They are carrying empty baskets on their backs and are seeking young cedar from which they can take strips of bark, and dig roots.

Following brown, needle-carpeted deer trails, we pass Douglas firs, their trunks shadowed deep in bark, ancient maples on whose limbs ferns and mosses grow, young balsam and spruce, with apple green tips on dark green branches. Close to the escarpment of the mountain, where stand giant boulders encrusted with moss, lichen and falling ferns, the women refresh themselves by cupping spring water from a woodland pool. Sunlight sends flickering shafts down into the brown green world.

There is the smell of smoke from a fire; a little further along the path, two of the women's kinsmen are burning, with hot rocks, the inside of a half-finished canoe. The cedar for the canoe was felled, by chopping and burning, last year. The men at that time shaped it into a small canoe and left it for the winter. Now, when they have burnt out enough of the centre, they will take the canoe down to the Shingle Spit for finishing.

The men are still burning and chiselling when the two women return with their baskets filled with folded inner bark, and coils of cedar root. The outer bark of the cedar strips, which they also have in their baskets, they give to the two men to use as fuel for their fire.

It is a day or two later. We are at the Shingle Spit encampment. Most of the men have left in canoes to troll off Norris Rock. Distantly, there is the drumming of grouse. Thud, thud, thud! One of the cedar gatherers is beating her fibrous inner bark into a greater softness. She beats the outer thickness, then soaks it in water, then dries it. A day or two later we see her folding it and putting it away. When the time seems right—when the weather is not good for gathering or drying, or next winter in the big house—she will make it into strips and weave it into a mat, a basket, or a cape. The inner thickness she is also softening for use as a shredded bark skirt or an apron. The softest pieces she puts away to use as a crib lining.

The root digger has been working since early morning removing the outside tissue of the roots by drawing each through a cleft stick firmly stuck in the ground. We now see her splitting each root in two, then in two again. She hangs the split pieces on the branch of an ancient fir. Later we spy her coiling them up to store.

Throughout the summer the digger weaves these roots, and other roots that she has dug, into baskets. Sometimes she sits on a beach, using sea water pools as basins in which to soften the strands of processed root. Into each basket she is weaving magical help for the user. The coarser roots she makes into burden baskets, the finer ones into containers for berry picking or for water. The latter she waterproofs with aromatic gum from the buds of cottonwood trees.

At Heron Rocks, a short canoe ride away, there are cottonwood trees. Later in the year she and other women collect the cottonwood bark. The inner bark, the cambium, is a great delicacy. The outside bark they bend and stitch into containers.

A battered truck is parked on the cement ramp at Phipps Point. It is an early spring morning; the mudstone slabs of the Northumberland Formation are deeply covered with ridges of loose seaweed. Two men in rubber boots are shovelling this lush dampness—a rich mixture of many different kinds of orange, purple, and vivid green weed—up and into the cab of the truck. When they have enough, the truck will be parked outside the Co-op Store. The seaweed will then be sold, and delivered, to older gardeners who do not have the energy to shovel it themselves, as islanders once did, into garbage bags on the beach. Rotted down, the seaweed promises healthy nourishing vegetables in the year ahead.

There is no cement ramp but the beach looks almost the same. Bare-footed women in bark skirts are finding red laver seaweed in the seaweed piles and are carrying it in armfuls to the small canoe anchored on the tideline. Their voices ring across the water as they paddle back to camp.

The canoe grates on the Shingle Spit pebbles. The women alight. One of them carries on her hip a burden basket full of clams. With much chatter they lay out the seaweed on the ground above the high water mark. A few days later, when it is dried, they put it away in cedar boxes. It will be used later for nibbling or for seasoning.

Flowers of the blackberry bush.

Chapter Thirteen

EARLY SUMMER

THE LEGACY OF THE GLACIERS IS SLIGHT IN Helliwell Park, and on all the headlands with southern aspects. On the windswept bluffs of St. John's Point Peninsular, Gabriola rock pushes through the grass, and joins with small glacial boulders to create rock gardens. Cactus grows here and, in the spring and early summer, there are wild flowers in profusion.

In the woods behind the bluffs grow berberis and evergreen huckleberry, as well as juniper, yew, oak, cherry, holly, arbutus and slow-growing fir trees. There is some water, and bordering the small swamps of broad-leaved sedges, grow willow and cottonwood trees.

The glaciers have left pebbles in a small bay; some will be pebbles wrenched by rasping ice from the Gabriola conglomerate. Today, rock hounds collect these pebbles to polish as gems. The people used them to create tools, and as boiling stones and fire stones.

The flowers are out on St. John's Point. Walking along the grassy headland is an older woman. There are leather moccasins in her basket; she wore these as protection from the spikes of cactus, as she clambered up the bank from her canoe. Also in the basket are chocolate lily and wild onion bulbs. She is collecting these to steam in a pit. The warm sun is bringing out the smell of guano from the comorant rookery on the cliff to our right.

There are patches of pink sea blush, splashed dark blue with larkspur, and dotted with clumps of yellow woolly sun-flowers. She walks past stretches of fading blue camas flowers. The camas bulbs will be good for eating later in the year. The woman stops to gently touch some violets, nestling in matted grass under a thicket of young fir.

Looking across the glittering water to Flora Island and on to Lasqueti Island, we see a pod of killer whales diving and surfacing across the Gulf. Men in canoes are trolling for ling cod, skate, rock fish or red snapper. They may not get many fish today with the whales around. Someone has beached his canoe on Flora Island, probably to look for seabird eggs.

It is the same time of year but in the twenty-first century. We are up on the lower shoulders of the mountain with local naturalists. Islanders always thought that the whole mountain was Crown land, belonging to them in perpetuity. But now it is realized that some of it is owned and is up for sale. Its future uncertain. The rest, once Crown land, has been re-designated "Groundwater Recharge Area/ Sustainable Ecosystem Management Area." Residents hope it will be preserved safely for future generations of Hornby Islanders.

Through the slim trunks of young Douglas fir, we glimpse blue sky. At our feet are mosses and lichens of all kinds, small salal bushes, young sword ferns, their fresh green fronds pushing up from brown crowns. One of the group points out selaginella, a plant resembling moss, but more closely related to fern, and nemophylium, a moss that resembles a flowering plant. Further along the trail we discover alum root, lichens of many kinds, some red tipped, some silver and flaky. We pick a pink, succulent, spotted coral root orchid, and look for its relation, the rattlesnake plantain.

Down the cliff of conglomerate rock to our left—we have to lie on our stomachs in order to see—is a vertical tapestry of deep yellow mimulus, red Indian paint brush, pink sea blush, and the tiny flowers of blue-eyed Mary. Below this the land flattens out into a bench. Cushions of dark green arbutus bunched with white flowers; carpets of soft yellow, the tops of burgeoning maple and oak. Immediately beyond and below again is the Lambert Channel. The water over rocks and the water over sand is creating blotches of dark blue, blotches of turquoise.

Mesmerized by the sea blues, we do not immediately realize that we have switched back into cyclical time. We hear the murmur of women's voices. People are coming up the trail collecting herbs. They have their baskets full of lichens, mosses, roots and leaves. Some may be made up into herbal potions and poultices to cure natural illnesses, such as head colds, boils, pneumonia, and wasp stings. Others may be used for what we might call "magical" purposes. One woman among the group has more authority than the rest. She may be a herbalist—a specialist whose knowledge is an inheritance given to her by her elders, supported by her sense of spirit power.

Throughout the summer we see women and some men from the Pentlatch families collecting roots, leaves and flowers. Plants that we understand from the hammerstone will help to cure sickness, ward off bad spirits, flavour food, create sweet smells. Everyone, it seems, has some knowledge of herbs.

In the Shingle Spit camp, in a special shelter erected for the purpose, there is a young woman, clasping a stick and pressing down. She is having a baby. Older women in the family are helping her, making sure that following the birth, certain rituals are adhered to.

It is another day. A young girl, at the start of her menses, is being secluded in a different shelter. Around her wrists, elbows, ankles and calves there are cedar bark bands. A ritualist is bathing her daily, chanting runes and incantations, as she paints the girl's face and sprinkles the parting in her hair with red ochre. She braids the girl's hair with goat's wool, sprinkles it with down.

At the end of the seclusion, there is a celebration. Special people have been invited. Gifts are being given. The passage of change has been safely negotiated. Life has changed for her; she needs from now on to be especially well behaved, hardworking, modest and virtuous, if she is to make a good marriage.

Here and there on the island, on the edges of the forest, in places where in recent times there was a fire, women are gathering berries. First salmonberries, then thimbleberries, then, as the seasons change, blackberries and huckleberries. The berries that they do not eat immediately are being dried back in camp, in the sun, and then packed into wooden containers. In the winter the fruit will be made soft again and eaten with the addition of oil.

The weather is changeable. The wind from the southeast howls, the rain pours down and the sea dashes against the rocks. For several days it is impossible to take out canoes, to gather berries or roots, or to do any drying of fish or fruit. Fires hiss and limbs start to ache.

Around the island people are sheltering from storms, either in their temporary house of bark and planks at the Shingle Spit, in their portable tents of matting, or under the spreading boughs of cedar trees. As they shelter they occupy themselves, chipping stone tools, dipping fishing lines into alder dye to make them invisible under water, stitching skin, weaving mats, keeping fires alight. Some young men are passing the time with gambling games. One old man is recounting a story:

"Long, long ago, when men and animals talked together and when animals were people, and people were animals, there was a problem. The south wind would not stop blowing. The north wind, who was everyone's friend, waited in vain for her turn to blow.

"So the animals and the birds, and the people and the fish all gathered together to decide what to do.

'We must not be angry with her,' one said, 'for we need her for rain.'

'We must reason with her,' said another, 'perhaps she does not realize what she is doing to us by blowing all the time.'

"They decided to send a spokesman. Not the bear, because he was unpredictable; not the eagle, he was too arrogant. Eventually, they decided on the skate, a very strange creature, for he is both very flat and very thin. Sometimes he is there, and sometimes he isn't.

"So the animals and the people sent the skate to be their spokesman, and because he could see both sides of the question, and because he knew that there was no one answer, the south wind listened to him.

"And this is why the south wind only blows sometimes, and the north wind has a chance, at other times, to bring good weather."

The skies have cleared, the sun is out. We are beside the massive stump of an ancient cedar. It is crumbling under mosses. Sprays of huckleberry with delicate, pale pink leaves are growing out of it. Around its base are traces of midden. For how many hundreds of years did people shelter here from the summer storms? What stories were told? What industry took place? What instruction went on?

What was it like to cut down this tree? Where did the planks of sawn cedar eventually go? Or was it split into shakes to roof a pioneer's house?

Chapter Fourteen

SUMMER

THE LAND SHADOWING THE NORTHERN BEACHES *is different from that at the Shingle Spit or St. John's Point.*

To the immediate west of Tralee Point the sandy/clay-like soil is the legacy of the sea when it retreated ten thousand years ago. Under a moist bank, stinging nettles and foxgloves grow. The sandstone beach yields a different intertidal life. Abalone, rock oysters and sea urchins, until recently, could be found at low tide.

The bay, where the petroglyphs are, is sheltered from the winds, though the northern beach in general is vulnerable to wind, weather, and to periodic invasion by strangers. It is near this place in the twentieth century that we found the hammerstone. It is believed that during the nineteenth century people from the north rested and watered near the petroglyphs on their canoe journeys to and from the white settlement of Victoria.

Shafts of sunlight are picking out patches of wet pebbles. In cyclical time on the northern beach just west of Tralee Point, a woman and her daughter, both with cedar bark blankets over their shoulders, for the wind is from the north, crouch near the edge of the water. They are using their digging sticks and their hands to find clams among the cold sand and pebbles. Their black braids bob and swing as they place pebbles on one side and clams in a rough basket on the other. Each time they dig, the hole sucks in water and sand.

On a nearby raised sandstone ledge, a man and a boy squat beside a canoe. One is tying cherry bark round a sinker stone, while the other is lashing a bone hook to a line of nettle fibre. They are about to go fishing.

Their camp of two shelters of matting on a grassy patch beneath a maple tree seems settled. These people may have been here since the day in the spring, when herring were trapped in the nearby stone fish traps.

Close by is a stream. It spills out of a deep gully on to a flat platform of sandstone. Into this sandstone, over the years, fishermen and sea mammal hunters, waiting for auspicious weather, have carved petroglyphs of fish, whales, and other creatures. The place is marked by a large boulder. Out of the wind, it is a place of sanctuary; with the rock carvings, a place with special spirit power.

Later in the day, the woman places clams in their shells on red hot coals and covers them with damp seaweed. The man and boy are out in the gulf trolling. The girl is further down the beach gathering kelp bulbs and stipes for storing oil.

The clams steam open. The woman prises a clam from its shell with a knife, and threads it on a stick of ocean spray. She then takes another. The empty shells she piles up neatly one inside the other and stacks them outside the fire pit. The several sticks of clams are barbequed over the fire. When the clams are cooked, the sticks are bent over and stacked. They will be dried further before they are stored for winter eating.

The shadows are growing long on the beach when the fishermen return. As they pull the canoe up on to the sandstone, the girl runs to help them. They have a large catch of salmon and rock cod.

The woman now has sea urchin, abalone and rock oysters sizzling on the open fire. Laid out on large maple leaves are pickings of early blackberries. In a bucket of cottonwood bark, she is making a hot drink. Strawberry leaves are being steeped in water made hot from heated stones put earlier inside the bucket. The fishermen are laying out their gutted fish on the grass.

Over on the other side of the island, at the Shingle Spit encampment, the headman is not feeling well. He does not want to consult the herbalist, who will give herbal potions and poultices, or the clairvoyant, who will perform exorcisms and recommend bathing. Nor does he feel he needs the shaman whom he would only invite if he was extremely ill and feared the sickness of his soul. All of them are his relations; he knows they have special knowledge and special powers but he is attempting to cure himself.

With the help of his wife he has heated up hot rocks within a sweat shelter on the far side of the camp. We see the wife carrying water in a bark bucket into the shelter. Then the headman, thoroughly hot from the steam, comes out and plunges into the sea. He returns to the shelter. When he has done this a number of times, we watch him vigorously scrubbing himself with cedar boughs.

Later that day we notice him receiving, with decorum and dignity, strangers who arrive in canoes from the north bearing gifts of oolichan oil. The "gift" is part of the economic system that makes everything work. It is given both to repay an indebtedness and to create an indebtedness.

The Coast Salish have a complicated system of give and take on several different levels, resulting in a balance of wealth. Through marriage agreements, through feasts, through potlatches, through times of unexpected bounty, people give to others within their vast extended family. By so doing, they gain prestige and also capital. Those who receive largesse know that they will need to give an equivalent gift back at some time.

During the 1980s, a system of indirect trading was tried on Hornby and flourished for a while. Known as LETS (Local Exchange and Trading System), it used its own dollars. I sell you some pots I have made, and with the LETS dollars earned, get someone else to clean my windows. The person who cleans my windows, is able to purchase LETS vegetables. A voluntary accountant made sure that LETS dollars balanced. Needs were met; no money profit was made. Everyone taking part searched for what they could give. The island seemed richer.

Today there seems to be universal acceptance that money exchange is the obvious token for survival, though it is true that trading, sharing and giving do still happen. Time has become, as elsewhere in the western world, a commodity. Many jobs that were once undertaken by volunteers are now being paid for.

The Coast Salish system, built up over thousands of years, is multi-dimensional. Every gift is interwoven with social and spiritual connotations, with expectations of behaviour, with time-honoured ritual.

The man receiving the visitors from the north was chosen many years ago to be headman of his winter house for his qualities of leadership, for his knowledge of the rights of others and for his understanding of traditional behaviour. While within the family he is soft-spoken, generous, and fair; outside the family his role has to be one of arrogance, conceit and ruthless detachment. It is he who is responsible for the allocation of fishing and gathering areas, for the organizing of potlatches, for keeping in balance the family's social and economic network.

On the other side of the Shingle Spit, while the headman is attending to his duties, a young mother is sitting, crooning and nursing her baby. The orange branch of an arbutus tree shadows her head. Laid out around her on slumped boulders of conglomerate rock are mosses. She is drying them to use as diapers.

Through the baby's nose and ears are threaded pieces of wool. These are to establish holes, through which when grown, he will—on ceremonial occasions—insert bone, wood or shell ornaments. The name the baby has been given is only temporary. When he is ten months old there will be a special naming ceremony; the name he is given then will carry with it certain privileges, rights, duties and possessions.

Beyond the Shingle Spit to the south, near Norris Rock, there are several canoes with men fishing from them. The men periodically call to each other.

In one canoe there is an older man and a young boy. The boy is being taught how to lash the bone point on to the wooden hook, and how to attach the hook to the nettle fibre lead, and then how to splice the nettle fibre line to a main line of kelp. The secrets of fishing in these waters are being passed on.

There is a jerk on the line. Very carefully, aware perhaps of the privilege of being allowed to hold the line, the boy pulls, lets go, pulls, lets go. He is bringing the fish slowly up to the side of the canoe. Deftly the older man shelves his paddle, and thrusting a spear through the side of the fish, flips it aboard the canoe. With a small club, the boy hits the fish on the head. Fresh bait is taken from a bentwood box, fastened to the hook, and fishing with instruction carries on.

Throughout our time of tuning in with the people on the island we see older people training and educating the younger. Knowledge is power. More than physical harm, the people fear censure from others in the family and from other family groups, but also from the spirits of their ancestors and from the spirit world itself.

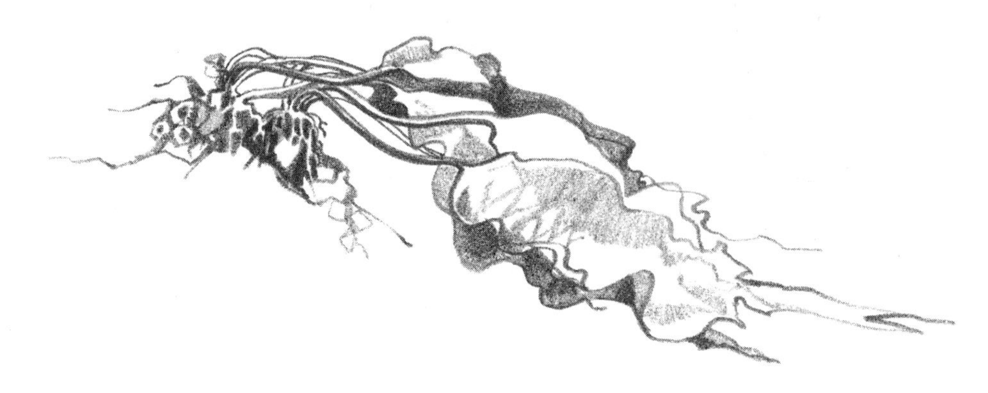

Kelp: seaweed uprooted and thrown ashore from deeper water.

Chapter Fifteen

LATE SUMMER

IT IS LOW TIDE AT WHALING STATION BAY—THE *large bay on the northern shore, a mile to the north of Helliwell Park, two miles to the east of Tralee Point.*

The expanse of white sand is coloured with summer people, pulling in boats, building sand castles, playing with frisbees, picnicking, sunbathing.

Sandflies hop in the sea wrack left by the high tide. An empty milk carton, perhaps washed in by the sea, is being picked up by an older resident as she tidies the shore. Along the edge of the beach, the windows of the summer cottages and houses are open. On the wind is the smell of outdoor cooking, someone is barbequeing . . .

. . . back in cyclical time, a whale has been washed up on Whaling Station Bay. News of the pending feast travels. Canoes have been plying the sun-splashed waters from the shores of Denman Island, from the Shingle Spit and from Ford's Cove, from St. John's Point and from Tralee Point. Men, women and children have left their immediate fishing or gathering tasks to share in the bounty. Whales are not hunted by the Salish, but when one comes ashore . . .

Everyone it seems has now arrived. By consensus, one man has been placed in charge. He is not the headman of the leading family, but the person thought best equipped to organize the party.

Some women are cutting up the carcass, others are lighting fires, heating rocks, gathering blackberries, cutting sticks of ocean spray to use as roasting skewers. Pieces of whale meat are soon sizzling on the fires. Excited talk continues.

A ripple of tension quietens the gathering. A new canoe is seen approaching the sandy beach. In it is a man wearing a cap of otter skin. There are feathers in his hair and the lower part of his face is painted red.

The Shaman! He and his assistant paddle the canoe hard on to the beach, and alight. As they walk up the beach, people back away, fearful of getting in his shadow, but the man in charge of the disposal of the carcass, welcomes the Shaman. He and his assistant are given a place beside one of the fires. A choice piece of sizzling meat is offered to him from one of the skewers.

The Shaman is a powerful person. He, more than anyone else, is in touch with the supernatural magic that is everywhere. His power is both respected and feared. He can see back in time, and into the future; he can go up into the sky and down into the earth; he can grasp souls and communicate with ghosts, and he can provide spells and cure sickness of the soul. He is a power for good, but should something go wrong for him, he can also be a power for evil.

Unlike the headman, he has not been chosen or elected. He has become a shaman by revelation, followed by vigorous ordeals. He had no formal initiation. His assistant may also become a shaman, but he will not reveal this aspiration until he can prove his powers.

Eating is now over and spontaneous dancing has begun. Stories of recent adventures are being acted out in mime. There is laughter, drumming, and games being played on the sands. Eagles, ravens and turkey vultures join in the feast. They are picking up bits left by the people, and flying to the tops of tall trees to nibble and watch all that is going on.

Darkness descends, warm air envelops. The wind is blowing off the land. The people sleep. Falling stars dart across the silvered sky. As dawn comes, the black range of coast mountains is etched red. A flock of crows flies low over the still darkened beach.

We are back in our time, waking up in the soft light of dawn, on the petroglyph beach near where the hammerstone was found. We clamber out of our sleeping bags. The low angle of the sun, throwing long shadows, is accentuating in the sandstone the carved outlines of two killer whales.

A man is crouching beside them. He has his face painted, his body oiled and eagle down on his hair. Something has changed. The cabin has gone. We step into the shadow of the trees.

The man is singing and, with a sharp stone, recarving the outline of the two whales. Two men with clubs stand nearby. Alongside them is a canoe from the bows of which juts a trident-like harpoon. The men are seal hunters. The man singing and carving is the harpooner. He is asking his guardian spirit, the killer whale to come out of the sandstone rock to help him in the hunt.

The incoming tide tosses the front of the canoe. With vigour the harpooner sings his song. He pecks the carving on the downbeats of the song, grinds the carving in the space between. His companions keep time thumping with their clubs. Some powdered ochre has been placed in a natural rock bowl near the canoe. As the tide floods the bowl, the men jump into the canoe, and push it off into redstained water. Gently they paddle north, into the wind.

Special paddles cut sound as the canoe glides against the ripples. A seal is sighted. They head towards him. Sensing danger, he dives.

Back in camp, a little to the west of the carvings, we see a woman, wrapped in a cedar bark blanket, staring out to sea, waiting. It is the harpooner's wife. While the hunter is away, she must not clean house, cook, or even build a fire, or comb her hair. If she is lively, the hunted seal will be lively too.

The hunters see another seal, his snout pointing away from them. They move closer. The harpooner in the bows gives silent instructions to the paddlers. He lifts his harpoon. The seal is two canoe lengths away. Then one and a half canoe lengths away. The harpoon is attached to a rope. The harpooner whirls the shaft through the air, and strikes. The harpoon hits. Multiple barbs enter flesh. Pulling on the harpoon rope the men play the struggling seal towards the canoe. A quick hit on the head and the seal is dead.

Back at camp the wife of the harpooner, now released from her confinement, helps cut up the rich seal meat. The liver is roasted for supper.

Over the next few days some of the meat is dried, some given away. The blubber is rendered down and poured as oil into kelp bulb containers. The men scrape the seal skin, so that it can be made into quivers for arrows. Nothing is wasted.

Chapter Sixteen

EARLY FALL

IT IS NIGHT TIME IN LATE SUMMER. PEOPLE ARE CAMPING on the St. John's Point Peninsular, close to Flora Island waiting for shoals of Coho salmon to swim up the Gulf.

Illuminated in silver are the matting shelters of the people. Silvered canoes rest on the foreshore etched in black. There is the soft orange glow of a fire, and the smell of burnt clam shells, aromatic herbs, fish. A sweet scent wafts from where camas bulbs, thistle roots, and the rhizomes of wild clover are steaming in a fire pit.

A heron squawks, disturbing the night. A shadow rises; a man places more bark on the fire. An owl hoots. Ghosts of the departed seem all around, spirits are everywhere.

While waiting for the salmon to come, the rhythm of the people's lives spirals on. Two women leave for the day to gather salal berries, oregon grapes, and huckleberries. Another woman is away searching for hemlock bark, and for the inner bark of the oregon grape from which she can extract dyes. Another woman—is she the one we saw earlier processing bark?—is collecting fireweed, thistle and cattail fluff to add to the weaving of her blankets and capes.

When the berry gatherers return, they crush the fruit, and spread the resulting mush out on skunk cabbage leaves to dry. The sun draws out the juices. Children coming back from the woods with armfuls of bedding—sword fern, bracken and hemlock boughs—drop their loads and through hollow stems of fat grass, happily suck at some of the sweetness.

A few days later, when the berries are dry, we see the berry cakes being stored away in bentwood boxes.

A woman is making matting from tule gathered in the nearby swamp. She is laying the stems side by side and sewing them together with twine threaded through a long bone needle, top to bottom, then bottom to top.

Men are making and repairing nets; they are flaking, chipping, abrading cobbles; they are splintering bone for fish hooks. One man is fashioning spears by inserting sharp stone points into stout sticks of crabapple. He uses nettle twine to lash the two together, and then covers the joint with pitch and strips of cherry bark.

Two men back from the forest carry lengths of ocean spray and crab-apple. These they tie up with cedar bark twine into bundles, ready for next winter when the lengths will be fashioned into shafts for spears, or harpoons, or made into digging sticks or mat needles. Some lengths may become gambling sticks, implement handles, dipnet poles, or even spoons.

In the evenings around the fires the people exchange news, gamble, joke, boast, and tease.

There are several families here, all related, either by direct descent or by marriage. Some "in-laws" may have been born to families living some distance away. Cousins can never marry and, for those nobly born, marriages are arranged long before puberty. Marriage is an instrument for cementing friendships between families, and through the exchange gifts, a part of the economic system. In addition to material goods, privileges, such as a family crest, or the right to certain spiritually potent songs and dances, are exchanged with the understanding that these privileges will be eventually handed onto the children. Unlike the Salish to the south, the Pentlatch suitor does not have to sit outside his potential bride's door for several days before he is officially accepted and taken in. Sometimes it is the man that goes to live with his wife's family. Sometimes it is the woman who has to go away.

If the salmon do not come soon, magic will be tried. Over at Tralee

Point, we see a fisherman re-pecking and grinding a stone carving of a salmon. He is singing his song to it.

Seagulls are seen gathering far down the Gulf. One of the fishermen, out in a canoe, paddles speedily in. The salmon are coming!

Salmon are caught here every year with reef nets above the kelp beds. In preparation the men have cut channels in the kelp, and have the nets, made of nettle fibre, and attached to anchor stones, rolled ready to place, between two canoes, for when the salmon arrive.

There is no flurry. Men appear, it seems casually, from the forest, the beach, or from the water. They push out and clamber into two long canoes. Each canoe is manoeuvred so that the stern faces the incoming tidal current. The net strung between the canoes is dropped and anchored. It is positioned so that the salmon entering the kelp bed swim into it from the cleared channel.

The women on shore watch silently. The sounds of the wind, the lapping water, and the cries of following seagulls are punctuated by the dropping of the anchor stones and the heavy breathing of the men.

The water is clear. We see the salmon in their silvered hundreds, flowing into the net. The leader in the off shore canoe raises his hand. A heave and then another heave, and the net is emptied. With a tumultuous wet rumble, a cascade of fighting salmon falls into the canoe on the near shore side.

All is now shouting and excitement. The women run down to the water's edge, and along with the men, club the heads of the heaving salmon. The fish are then carried one by one up the grassy bank.

They wipe the slippery scales with mosses and grasses, slice off the heads, tails and fins with sharp mussel shell knives, slit the undersides open and remove the innards. Finally they peg down the filleted salmon with cedar splints on racks to dry.

For the next few days the people feast on the heads, tails, fins and bones of the salmon, roasted or boiled. All that that they do not eat, they reverently return to the sea. They believe that returning the parts will regenerate the whole fish.

Flowers of Ocean Spray (Holodiscus Discolor) commonly seen in June and July on Hornby Island.

Chapter Seventeen

FALL

STRACHAN VALLEY IS U-SHAPED, AND ABOUT HALF *a mile in length. It nestles under the protection of Mount Geoffrey to the southwest, and is sheltered by a low ridge of land, part of the sloping Geoffrey conglomerate, to the northeast. For thousands of years a swampy lake, it was drained by early settlers, and throughout the twentieth century provided farm land and rough pasture. Now in the twenty-first century, thanks to the action of beavers building a dam, part of it has again become a lake. As a result of this, the island water table is rising, and sweet clear water flows once again down Ford Creek to Downes Point and the sea.*

In cyclical time, seen from the grass mound at the top of the mountain, the valley is defined by clumps of cedars and by lake and swampy meadow land.

A small party of men, women and children are making their way, laden with baskets of all kinds, up the Ford's Creek to Strachan Valley.

The woods around the lake and swamp hang heavy with portent, for this is a holy place, and, in the past when danger from raiders has threatened, it has been a place in which to hide.

Besides baskets, the people are carrying nets and clubs for catching grouse, knives of bone, slate and mussel for cutting rushes, mats for shelter, crushed dried seaweed for seasoning, and a willow root spindle with drill, to be whirled between the palms of the hands into a cedar hearth for making the initial flame for a fire.

As they set up their camp, on a mound close to the woods overlooking the swamp, grouse—as plentiful as domesticated chickens,—peck the berries off the salal bushes. A young doe casually browses on late blackberries. The next day two men take off to hunt and the women start to cut the marsh sedges and to pick the crab apples, which, though still green, are full size and can be ripened in storage.

The two men clamber up a trail to the ridge above the valley. Each man carries a bow of yew, strung with sinew. Around their waists are quivers, made of seal skin and cedar bark, holding cedar arrows. These arrows are about the length of an arm and are finished with two eagle feathers at one end, and with arrow points of slate at the other. The points have been shafted in such a way that when an animal is pierced, the arrow will drop off, the point remaining in the wound.

On top of the ridge, maples have changed colour, and are standing in pools of their own gold, brightening the forest. They see the buck before we do. It flinches with the first arrow. Buttocks bouncing, white tail fringed with black, it crashes into deep salal. The second arrow hits it. The deer crumples.

The older of the two men takes his knife of sharp obsidian, and thrusts it through the tough, rough skin of the deer's neck. Then drawing the knife across, he slits the deer's throat. Dark red blood gushes out. He lets the blood run over his fingers, then flicks it in four directions. This is to appease the ghosts that are watching.

Being careful not to step over the deer, they lash the two front and the two back legs together with rope made from cedar, hitch the deer on to the younger man's shoulders and take off, not in the direction of Strachan valley but down the other side of the ridge to Ford's Cove.

There has been a family camped at Ford's Cove all summer. At the Cove there is shelter from the wind, sweet water, easy access to good fishing and berry picking areas. As well, a swamp nearby grows fibres with which to make mats and capes. And in the fall there are acorns to pick and camas to dig at nearby Heron Rocks. It is now late fall.

The two men come back from the mountain with the deer and we see them slitting it down the stomach and pulling off the skin, in the same way that we would pull off a sweater. The entrails they cannot eat, they bury. Two women take over, jointing the animal, and cutting the meat into strips for drying. Bones that they cannot make into implements, they put into the sea. They exhibit care and a reverence for what they are doing.

While the deer is being cut up, an older woman is seeing to her cooking. She has a fire burning at the bottom of a pit lined with stones. Two young women who have just returned from Heron Rocks give her a basket of camas bulbs.

The woman with the steam pit removes the ashes and unburned wood from her pit, and places damp seaweed, leaves, and lichens on the hot rocks. On this she scatters some clams, then she layers on more seaweed. Next, with the help of the younger woman, she wipes the camas bulbs free of dirt, and spreads them out in the pit. All three then add more seaweed.

They scrape up soil from the forest bottom nearby, and dump it on the last layer of seaweed. Finally they lay old matting over the pit, and push the hollow stalk of a salmonberry through the matting. During the next few days they keep pouring water into the stalk.

The people from Strachan Valley return some days later, in time to taste the camas. It is soft brown and sweet, and also, because of the clams, a little fishy.

Outside a hut of cedar bark with a cedar bark roof, sits the grandmother of the family. She plucks the grouse that were snared in Strachan Valley. She is carefully putting the down in a pile. It will be spun later into yarn for weaving. The grouse are roasted in the evening on a spit, and the bones are carefully kept to be made into needles, pins or sharp points for fish hooks.

Fishermen return in their canoes, young boys dive and swim, the pebbles beneath them rippling with sunlight. Last-minute herbs, roots, and berries are gathered. Soon, the people will take down their shelters and return to their winter villages.

Mists hang over the fields. Golden sunlight sends long shadows from the boulders on the beaches. There is a smell of moist decay from autumn leaves. Most of the summertime residents have gone, some of the long-time islanders, those with children going to high school, have also left. Perhaps they will come back to the island when their children have graduated.

At the Co-op Store there are new faces, young people who in the summer came to the island, found they liked it here, and stayed—single mothers, looking for a safe place to bring up children, young men looking for new meaning in their lives. Other new faces are those of older people, newly retired, perhaps one-time summer visitors, who realizing a dream, have now turned their summer cottages into year-round dwellings, and become part of the changing community.

Longtime residents are canning or freezing tomatoes, pulling up and composting the dead haulms of peas and beans, picking apples, pears, making wine from blackberries. Deer with spotted fawns still tiptoe across the road from wild land to wild land. Islanders are beginning to again "see" islanders, sit on committees, sign up for courses organized by the Community School, proferred by islanders with skills or special knowledge. With late-night ferries only on Fridays, islanders have to find evening entertainment and instruction from their own resources. The vitality on the island may seem different to that of twelve years ago, but it is thriving and creative.

The first people do not stay on the island through the winter. They are moving back to their winter villages. At Ford's Cove as well as at the Shingle Spit, and over on the north shore, canoes of several sizes are being stacked. On them are mats and cedar planks; cotton wood bark containers filled with dried salmon and dried clams; baskets full of dried berries; bundles of coiled roots, cedar bark, and stripped nettle stalks; lengths of dried tule, sedge, and bark of all kinds for making dye; pieces of hemlock, maple, alder yew, oak for carving; baskets of bones, for fashioning into knives, needles, ornaments, spear heads, fish barbs; rocks of all kinds, relics of Hornby's far distant history.

Some of the bounty of the island has already been carried back to the winter village, for the people are not entirely divorced during the summer from their winter home. Bad weather, illness, the need to dispose of some of the "gatherings", or a family occasion have demanded periodic return. We may also see on the canoes shell ornaments, fine baskets, bows of crab apple and yew, and arrows of cedar, blankets of cedar bark softened with fireweed and thistle fluff, platters of alder, spoons of maple, combs of hemlock, and fine new strips of cedar planking.

In between their gathering and preserving, the women will have found time to create this wealth. And wealth it is, for these are goods that can be used in a potlatch, either this year or in some year to come. We see the canoes go, and then come back, and go again. Not everything is carried in one journey.

Pock-marked rocks at Spray Point.

Chapter Eighteen

WINTER

THE ISLAND MOVES INTO WINTER. THERE ARE DAYS WHEN clouds move in and brown-south easters blow; days when it never stops raining; days when the winter sun warms the trunks of trees and the mainland mountains float, white and silver, above pale clouds; days when deer, in search of salt, tiptoe on the frosty beach, and days when thick snow reveals in footprints the continuing life of the island—the dragging trail left by a beaver's tail, the doglike prints of the otter family, the two pronged tracks of deer, the delicate imprints of birds.

One day, not long after the people have left, there is a succession of canoes slowly paddling around from Chrome Island over to Downes Point. One of the elders in the winter village has died. His body is in one of the canoes, flexed in a womb position, wrapped in a cedar blanket and inside a cedar box. He is being taken to one of the caves at Downes Point. The box will not be left there for ever. After a year his bones will be removed and buried elsewhere.

The people are still not far away. Men come to the island, sometimes staying for a few nights. In some years solitary families may spend winters on Hornby. And in other years the Shingle Spit may be used as the site of a small winter village.

When the foundations for the pub at the Shingle Spit were being dug, a number of human bones and artifacts were dug up, including a large stone bowl. Stone bowls were not for every-day use, but were used by shamans. Soon after the bowl's arrival in the twentieth century, it disappeared. It was left outside the door of the pub owner's house. Did the person who took it realize its value? Or? . . . Some say the Shingle Spit was a "holy place," and development should never have taken place there.

On one moonless night, there is light on the dark water of the Lambert Channel. Resting duck are being dazzled by the light of cedar bark torches in the bows of a canoe. Seeking to escape, the ducks paddle into the shadow of the stern, only to find themselves caught in nets. The nets are lifted, the ducks knocked on the head and thrown into the belly of the canoe.

When the tides are low, bark torches illuminate people digging for clams.

We are on the island, but it is difficult to say where because the island is enveloped in white mist. This mist has both silence and sound. It is packaging, in stillness, every living thing. No movement. Just the thinning and thickening of cloud.

The sound of a wedge being hammered by a stone into a log rings clear. The silence closes in. As if harsh silk is being torn, a cedar log is wrenched in two. Again silence. The veil of mist thins, bringing into sight a line of firs, pale silhouettes. Then these too fade.

The hammering has stopped. The mist closes in. Silence.

Chapter Nineteen

HISTORICAL TIME: 1790 A.D. TO PRESENT

THE SPANIARDS, SURVEYING TEXADA AND LASQUETI IN 1791, step on the southern shores of Hornby Island, and seeing her beauty, proclaim her Isla de Larena, the Island of the Queen.

As if a bell has been rung, this is the start of the end of Hornby having people as part of her environment, ebbing and flowing with the seasons, complementing her bounty, one with her cyclical nature.

The people in the winter villages do not know from where "The Sickness" comes. Every year, more people die. Herbalists try new remedies, shamans extend their powers, rituals are severely enforced. But still people die.

Year by year fewer people come to Hornby. The birds eat some berries, the rest fall uneaten to the ground. The campsites grow lush with untrodden grass.

Boats with white sails are seen sailing up to Quadra Island. Large canoes are seen speeding south.

Diseases come ahead of the traders. With the traders come metal for speeding woodworking skills, firearms for more efficient hunting and fighting, and a new way of acquiring wealth that has nothing in common with the ancient concept that wealth is a reward for virtue and fair dealing.

A different kind of man now becomes wealthy, and because wealth is equated with leadership, a new kind of man becomes leader.

Trading does not touch the Pentlatch directly. Trading to them meant firearms in the hands of others, and new codes of behaviour from men who have larger and larger war canoes.

The Haida and the Tsimshian come south in search of slaves and take the Pentlatch. The Kwakiutl, with firearms from the Nootka, push south, acquiring Quadra Island and Campbell River from the Comox. The Comox in turn push south into the land of the Pentlatch.

Hornby is no longer the exclusive hunting-gathering-fishing ground for a traditional group of people. It is accessible to everyone who comes this way.

Fear is everywhere, fear not of the white man, but fear of the new-found power of neighbours. Fear from the loss of family and of leaders, both headman and shamans, and from the apparent loss of spirit power.

Girls venturing to pick berries on Hornby hide their canoes and are escorted by an armed protector.

Fishermen paddling to the shelter of a familiar bay on Hornby are in danger of being attacked by visitors from the north, lying in ambush.

Fort Victoria increases in size as loaded canoes in their hundreds go up and down the gulf. The northern shore of Hornby is used as a watering place, and sometimes as a base camp for raids. Pentlatch, camping there for traditional fishing, build their fires up on the bank among the trees, to deflect attention.

Fires sweep through the Hornby forests from camp fires hastily deserted.

There are fierce battles spreading from Dunlop Point across Sandpiper to Downes Point, and up into Strachan Valley, and from the Shingle Spit across to Phipps Point. Mounds are built as fortifications in case of attack.

In the 1850s, a few remaining Pentlatch, camping on Chrome Island after hop picking in Oregon, are raided and wiped out. One elderly woman escapes by hiding under the bodies of her family.

Sickness continues. In the 1860s, the beaches of Hornby are littered with dead bodies, sometimes in piles of ten. They are smallpox victims who died while escaping from the epidemic in Victoria. They are left on the island by their companions travelling back north.

White settlers come to the island. There are reports in the *British Colonist* of the first settlers in 1863 having their house burnt down, reportedly by "Euclataws from Denman," and of a murder of one white man and two Indians over "an affair of selling whisky."

The settlers have a vision of farming. To plant the orchard at Ford's Cove, the forest has to be burned.

It is discovered that the by-product lumber from the clearing of land for farming brings in more money than the farming itself. Forests are destroyed on Hornby for the money the timber brings in. Trees are cut down wholesale, and what is not exportable, because of size, immovability, or a slump in the market—as happened to cedar in the early 1950s—are left to rot.

The skyline shrinks, shadows thrown by trees shorten, and the beaches are fringed thickly with timber from storm-broken log booms.

There is more sunshine on the land. Acres are turned into orchards, grazed and tilled as fields. The trails following ley lines first used by deer, and then by the first people, become roads for the settlers.

After the 1870 Land Survey, needed for the establishment of the legal ownership of land, roads follow property lines, and a grid system, like a net, symbolizing the attitudes of western civilization, starts to stretch over the dips and curves, bluffs and benches of Hornby.

Hornby, though, is not for capturing.

Land not constantly tended swiftly starts returning to bush, then forest.

The forces that have created Hornby over the past 350 million years are still at work. The music of the universe will still be playing long after man has gone.

In a silvery dusk at Whaling Station Bay, eagles tinkle their notes to a grey flocked sky while sandpipers scurry along the incoming tideline pecking at the pewter sands. Summer cottages darken into oblivion, trees grow taller.

Canoes are surely drawn up above the highwater mark, dark shapes are moving, and along the beach, isn't there a glow of fire?

I give you the hammerstone. We slip into geological time. Are we seeing the seas rising, Mount Geoffrey crumbling, Hornby Island still moving north?

Rondo

A BOOK THAT STARTS WITH THE CHAPTER HEADING "Prelude" should perhaps end with a chapter called "Finale." But this isn't the end. Everything is still "in the beginning." The island story goes on.

The molten centre of the earth is continuing to spew forth basalt under the seas to create moving oceanic plates, which, when meeting continental plates, are causing earthquakes, volcanoes, mountain building. The water, the soil, the air, the plants and we ourselves are in the constant state of being recycled. Vancouver Island is tilting, Mount Geoffrey is eroding. The sun rises, the moon sets, the tides come in and out. The rhythms of the earth, without any orchestration from us, carry on.

Under the warm, shadowed sandstone, there are still layers and layers of Cretaceous history waiting gently to be revealed by a million whittling seas.

More immediately, the past comes to greet us, there are middens waiting to reveal their secrets, and fossils that have not yet told their tale. In the soil are spore traces as yet unanalysed.

Two little boys, in shorts and t-shirts, are crouched on the beach, gazing into a deep pool of water left in a sandstone basin by the receding tide. Tiny crabs are scuttling, fingerlings darting in and out of waving seaweed. What other wonders of life are they seeing there?

Over towards Texada, a cruise ship, large, white, many-decked, is slowly moving north. Its passengers, amid all the entertainment offered to them, are able to view, should they look out on the port side, a green treed, cottage-dappled, outline of an island, Hornby.

High in the sky there is the white trail of a jet plane. World travellers are up there, perhaps on the move from Tokyo to Vancouver, or Alaska to Seattle.

On Hornby itself stretched out behind us, islanders are tapping on computers, activating electronic waves, sending/receiving information faster than the speed of light. They inhabit cyberspace time, not geological or seasonal cyclical time, not mythical time. This is a time related to the perhaps temporary grid of human busyness.

The beaver has returned to the pond in the woods near our house. We stand very still in the late evening; he is swimming in the dark water. A family of river otters has clambered up the gully to dive and slither over and under a floating log. In the fading dusk we see, near a clump of trees, a deer watching us. His large ears are fanned out. He is listening. Turning tail, he thumps into the darkening forest. An eagle, high in a tree, motionless, watches the scene.

Down near the cabin grasses are dying back after spilling their seed, blackberries are ripening, and on the shore the sea comes in and out, slowly whittling away at the rocks.

Driftwood logs still unpredictably roll down the sandstone beach and splash into the sea. Earth tremors are still present. An imminent earthquake in this area has been predicted. Islanders are now organized to cope, should it happen.

Many of the fears carried twelve years ago about the immediate future for the island are still with us: fears that enterprises will be opened up and money invested by people insensitive to what it is that the island has to offer; fears that we may allow "progress" to take the place of "balance"; that the uniqueness of the island, the specialness that grows in secret corners with its roots far far back into the past, may be lost to us.

Some of the fears are perhaps already being realized. Climate change from the greenhouse effect, and holes in the ozone layer, are already making more violent our normal weather patterns, the sea has yet to rise and our gardens yet to die. We hope they won't, not for hundreds of years.

While the number of people living here has not gone up, island demographics have changed. There is now a much larger proportion of older people than younger. Several parents of teenage children have found it necessary to move off the island in order to educate their children in the faster moving, contemporary world. School numbers have gone down.

Despite many urban values appearing to be brought on to the island, I like to believe there is a returning recognition of the need to see and feel all that the island has to give, and then rest in it.

"The rhythm of the cosmos is something we cannot get away from without bitterly impoverishing our lives," D.H. Lawrence wrote in the 1930s. "Man has little needs and deeper needs . . . we have fallen into the mistake of living from our little needs till we have almost lost our deeper needs in a sort of madness. . . ."

Change is inevitable, and not necessarily to be feared. The island can carry more people and not be temporarily destroyed. But if our children and our children's children are going to be able to experience fully the island's natural life, then all of us who know the island must think of ourselves as guardians and allow only changes that are in harmony with its natural music.

In 50 million years, it is possible that Hornby will again not be an island. It will undoubtedly have changed shape. Perhaps it will be flatter. It will be further north, and will carry different vegetation, shelter different creatures. Some of the grasses, trees, shrubs will be the descendants of plants we know.

I like to think that our descendants will still be here, too, not trying to capture and control the world in a man-made net, but freely dancing, hammerstones in hand, to the rhythms of the universe.

Salmonberry: an early fruit that is an example of the rich vegetation on the island.

Afterword

IT IS SPRING AGAIN ON HORNBY, AND THOUGH MANY WILL say, with truth, that Hornby, in the last twelve years, has changed, my Hornby, what I like to think is the real Hornby, is still here.

Last night there was the slither of a moon above tall dark trees, and the deafening cacophony of mating frogs. Earlier this morning, the two Canada geese, that each year bring up their young on our pond, flew over calling, in squeaky-wheel voices, "We are here! We are here!"

I dig into the rich soil of our vegetable garden, earth inherited by us as forest bottom, enriched over the years by seaweed, barn manure, and the periodic scratchings and droppings of dozens of hens. Near me, the dark green rosettes of early potatoes are already showing, and beyond, the pale green shoots of broad bean plants.

There are as many different visions of Hornby Island as there are people living here.

On my Hornby, and I like to think it is yours too, it is still possible to tune into the natural cycles of the earth and the seasons. It is possible to eat fruit and vegetables as they come into season, spend time digging, manuring, and planting, weeding, watering and harvesting. And to feel rich and rooted as a result.

It still does not seem to really matter if the lawn is not regularly mowed. The crocuses, scyllas, daisies, and dandelions, push through the growing grass in spring, giving unexpected delight. They are followed by daffodils. Yes, my husband John does complain that the dying leaves are untidy.

Acceptance of imperfection is part of Hornby living.

Sometimes crops fail, predators—mink, eagle, possum—take chickens, sheep get trapped, plants die, wells go dry, the store may not have exactly what you planned to purchase, ferry lineups are long and tedious, nothing is certain.

I start weeding the overwintered brassica—spring cauliflower and purple sprouting broccoli. Chickens on the other side of the wire fence noisily gobble the weedy greenery as I throw it over. Suddenly, silence, the silky sound of an eagle's wings as it, slowly, ominously, flies low above our heads. I go on digging. A wind from the west cools my cheek. The chickens start their chortling again.

Place Names

BAYNES SOUND Water between Denman and Vancouver Islands, 67.

BEAUFORT RANGE Mountain range shadowing the east coast of Vancouver Island from Comox Lake in the north to Horne Lake in the south, 17, 26, 67.

BELLA COOLA Town situated on mainland coast north of Vancouver Island; south of the Queen Charlottes, 61, 62, 63.

BUTTLE LAKE Lake, adjacent to Strathcona Park, north of Courtenay, about eighteen miles long. It flows into the Campbell River, 23.

CAMPBELL RIVER Fishing resort and logging town, north of Courtenay on Vancouver Island. In prehistoric times, it was the territory of the Comox Indians, 114.

CHROME ISLAND Small islet on southern end of Denman Island, 111, 114.

COAST RANGE Mountains stretching north along the mainland coast of British Columbia from the Fraser River to the Yukon, 17, 23, 27, 44, 48, 56, 58.

COLLISHAW POINT Northern extremity of Hornby Island. Area where many fossils are found, 41.

COLUMBIA RIVER River flowing from south east British Columbia, through Washington State, and out into the Pacific along the Oregon-Washington border. It is 1,214 miles long, 57.

COMOX Small town and harbour to the north of Hornby Island; its valley and estuary, into which the Courtenay River flows; and the name of the band of native people presently living in this area, 31, 55, 61, 64, 66, 70, 114.

COURTENAY City on Vancouver Island to the northwest of Hornby Island, 1, 61, 64.

CUMBERLAND Village, slightly inland on Vancouver Island, directly west of Hornby Island. One time a coal mining centre. In the Cumberland coal seams fossil traces of Cretaceous flora have been found, 30, 31.

DEEP BAY Fishing harbour on Vancouver Island, to the south west of Hornby Island. In its middens are traces of human habitation dating back 5,000 years, 61, 63, 64, 66.

DENMAN ISLAND A companion island to Hornby, it lies between Hornby and the east coast of Vancouver Island, 13, 17, 35, 36, 58, 67, 76, 97.

DOWNES POINT A privately owned, wooded peninsular, on the south western shore line of Hornby. In its caves, Indian relics have been found, 66, 105, 111, 114.

DUNLOP POINT A spur between Little Tribune Bay and Sandpiper beach, on the south eastern side of Hornby, 114.

FLORA ISLAND Small island off the tip of St. John's Point, 63, 86, 101.

FORD'S COVE A sheltered harbour on the southern shores of Hornby. Named after George Ford, one of the first settlers, 14, 33, 62, 70, 76, 97, 107, 109, 115.

FOSSIL BEACH Beach on the north west side of island, 39, 41, 80.

FRASER VALLEY Valley through which, for the last eigthy miles of its journey from the Rockies, the Fraser River flows. It spills out into a delta at Vancouver. The shifting habitat of the first people for thousands of years, 61, 62, 63.

GALIANO ISLAND Island on the outer edge of the southern Gulf Island chain. To a much greater extent than Hornby, it's rocks were deformed and tilted when, in the Eocene, the Juan de Fuca Plate added new oceanic terrane to the southern end of Vancouver Island, 32,43.

GALLEON BEACH Subdivision on Hornby Island's northern shore, 37.

GULF OF GEORGIA Stretch of sea between Vancouver Island and the mainland, beginning at the northern end of the San Juan Archipelago, and extending in a general north westerly direction up to Cape Mudge; a distance of 110 miles, 5, 23, 27, 49, 50, 51, 55, 66.

HELLIWELL PARK Provincial Park. Three hundred acres of scenic bluffs and forest, situated on south western side of St. John's Point Pensinsular, 33, 39, 47, 85, 97.

HERON ROCKS Oak grove and bay on the southern shores of Hornby, a short distance east of Ford's Cove. Presently the site of a co-operative camping ground, and of a Friendship Society, 33, 62, 76, 82, 107.

KITSILANO Residential waterfront area to the southwest of downtown Vancouver, 48.

KOMASS BLUFFS Sandy bluffs on north end of Denman Island, 55.

LAMBERT CHANNEL Stretch of water between Hornby and Denman Islands, 17, 33, 67, 75, 87, 112.

LASQUETI ISLAND Island in the Strait of Georgia to the southeast of Hornby. Its rocks are older than Hornby's being composed mainly of igneus rock of the Karmutsen Formation, 86.

MILLARD CREEK A creek, a little north of Royston on Vancouver Island, where several archaeological digs have been made, 61, 64.

MOUNT GEOFFREY Highest point on Hornby Island; height about 1,000 feet, 16, 17, 37, 50, 58, 62, 105, 116.

MOUNT ST. HELEN'S Volcanic mountain in the State of Washington that blew May 18th, 1980, and continued to blow intermittently until Sept. 7th,1981, 50.

NANAIMO City and port, about fifty miles south of Hornby Island; originally a coal mining town. In the coal seams have been found fossilized imprints of Cretaceous flora, 6, 25, 29, 31, 43.

NORRIS ROCK A sea gull nesting islet several hundred yards to the south east of Heron Rocks, 63, 81, 95.

OLSEN'S FARM Farm with southern aspect immediately to the east of Heron Rocks, 50.

PHIPPS POINT Sandstone spur on west coast of Hornby, about one mile north of the Shingle Spit, 35, 76, 82, 114.

PORT MCNEIL Town on eastern shore of northern Vancouver Island, 32.

QUEEN CHARLOTTE North of Vancouver Island, separated from the mainland by the Islands Hecate Strait. They share the Wrangellian history, 21, 25, 27.

QUADRA ISLAND Island north of Hornby, east of Campbell River, 54, 55, 66, 113, 114.

QUALICUM, BAY, BIG AND LITTLE RIVERS Places on east Coast of Vancouver Island, north of Qualicum Beach, and Nanaimo, south of Courtenay Name given to Salish Band who live there near the Big Qualicum River, 56, 70.

ROYSTON Town on east coast of Vancouver Island, north of Hornby, and a few miles south of Courtenay, 61, 64.

SANDPIPER Subdivision with shaly beach facing south east, located between Downes and Dunlop Points, 39, 62, 114.

SHINGLE SPIT Sandy spit protruding into the Lambert Channel on the west shore of Hornby. Near the ferry landing, 14, 35, 62, 64, 67, 68, 76, 80, 81, 83, 88, 91, 93, 94, 95, 97, 109, 111, 112, 114.

SPRAY POINT Spur of land between the two Tribune Bays, 39.

ST. JOHN'S POINT Most southerly tip of Hornby Island, and of Helliwell Provincial Park, 17, 33, 39, 43, 47, 85, 91, 97, 101.

STRACHAN VALLEY About a mile inland to the north east of Downes Pt. The south western side of Mount Geoffrey curves round it, and Ford's Creek flows out of it, 17, 18, 56, 105, 107, 114.

TEXADA ISLAND Island lying between Hornby and the Coast Range. Rocks mainly of Sicker and Karmutsen Formations, 22, 23, 58.

TRALEE POINT Treed spur on the northern shoreline of Hornby, 10, 58, 62, 64, 76, 91, 97, 102.

TRENT RIVER Flows into Comox Bay from Vancouver Island, lightly east of Royston, 30.

TRIBUNE BAY Major sandy beach, with a south eastern aspect down the Gulf, 14, 17, 39, 40, 66, 73, 74.

TREE ISLAND Small islet on the northern tip of Denman Island, 64.

TSABLE RIVER Flows into Baynes Sound from Vancouver Island, southwest of Denman Island, 30, 61.

WHALING STATION Large sandy beach on the northern shore of Hornby Bay, 97, 115.

WILLEMAR BLUFFS Bluffs to the northwest of Comox Bay, 55.

Glossary for Part I

THE TRAVELLING LAND

ALEXANDER TERRANE An exotic crustal fragment, consisting of a diverse assemblage of Paleozoic rocks (600-230 million years old) that moved north ahead of Wrangellia. Wrangellia collided with it about 140 million years ago. Its rocks are now to be found as part of Prince of Wales and Chichagof Islands, and the adjacent mainland.

AMMONITE An extinct type of squid like animal, encased in a chambered shell; usually coiled in a spiral. Consistency in the changing of the shell pattern has enabled paleontologists to date strata from the design of each ammonite fossil found.

ARGILLITE Compact sedimentary rock composed mainly of clay.

BASALT Dark, fine-grained igneous rock of a lava flow or intrusion.

BEARD, GRAHAM Runs the Vancouver island Paleontological Museum in Qualicum. He is a noted amateur paleontologist. Finder of the mosasaur skull on Fossil Beach, and of the walrus skeleton at Dashwood.

BRACHIOPOD A lamp shell. A mullusk like marine animal having a dorsal and ventral shell. Superficially resembles a clam or an oyster, but is not related.

BRYOZOANS Moss animal. Marine or freshwater animal which forms colonies of polyps. Their delicate preserved shapes look like mosses.

CACHE CREEK One of over a hundred terranes, all with different geological Terrane origins and histories, that make up land west of the Canadian Rockies. According to fusulinids (marine micro-fossils) of Permian origin found in its limestone sediments, this terrane originated 350 million years ago in the Tethys Sea, (today's Southwest Pacific Ocean). It was formed of oceanic crust to which a platform of basalt was added. As it moved north and east, it accumulated several thousand metres of limestone.

CHERT A hard compact rock consisting essentially of microcrystalline quartz. Flint is a variety of chert.

COBBLE Waterworn rounded stone; bigger than a pebble, smaller than a boulder.

CONCRETION A rounded mass of mineral matter occurring in sandstone. Often forms around a nucleus of organic origin.

CONTINENTAL PLATE Continental plates consist of relatively thick and buoyant crust and lithosphere. Unlike oceanic plates they are never plunged back into the centre of the earth. Much older than oceanic crust (parts of them are over 3.8 billion years old), their recycling is by erosion, through the agents of air and water, by accretion through volcanism and by tectonic action on their boundaries. See Tectonic Plates.

CRETACEOUS Period in the Mesozoic Era, lasting 70 million years, from 65 million to 135 million years ago. Warm conditions were world wide; angiosperms, flowering plants, appeared and insects evolved to collect the pollen of flowering plants.

CYCADS An ancient plant, in appearance both fern and palm tree. Some species have a thick unbranched trunk, topped with a crown of large leathery feather shaped leaves. Fossilized cycad leaves havc been found on Vancouver Island.

DE COURCY FORMATION First cycle of deposition of Cretaceous sediment to reach what is now Hornby Island. Laid down in a delta environment, it can be seen outcropping between Ford's Cove and Heron Rocks. Most of Denman Island is De Courcy Rock.

DENTALIUM A tooth-like shell. Found on Hornby in the Cretaceous Period. Traded from the Nootka, in prehistoric times, by the Coast Salish, for use as currency and for ornamentation.

DETRITUS Rock in small particles worn or broken away from the mass by action of water or glacial ice.

EARTH Outermost layer is the crust, then comes the mantle, and structure of finally the core. The crust is very thin. Oceanic crust is made of basalt; continental crust of granite. Continental crust is lighter in weight and much thicker than oceanic crust. It is also in some places over 3,500 million years old, whereas oceanic crust is seldom over 100 million years old. The outer rigid of the earth's mantle plus the overlying crust is known as the lithosphere, the lower part hotter, weaker and partly molten is known as the asthenosphere. Its consistency enables the rigid plates of the crust above to move about. The upper mantle is believed to consist of dense, dark coloured rocks, probably peridotite, and of iron and magnesium rich silicate minerals. Fragments of these rocks are found in magma. The core is twice as dense as the mantle. The outer core which is molten generates the earth's magnetic field. The inner core is thought to be solid.

EARTHQUAKES A series of shock waves generated at a point (focus) within the earth's crust or mantle. The point on the surface of the earth above the focus is called the epicentre. The principal cause of all earthquakes is the fracturing of rocks following the gradual accumulation of stress during plate movement. The severest earthquakes tend to occur at the plate margins, where plates slide past each other. Rock has elasticity and, where areas of rock are subjected to the push-pull forces of plate movement, stress gradually accumulates. When the strain becomes too great, the rock ruptures. Tension is released as the rocks on either side of the fault spring back violently. The energy released causes the ground to oscillate. It is these waves that enable scientists to pinpoint the epicentre of an earthquake and to measure its magnitude.

ECOSYSTEM The natural interaction of all things within an environment.

ELASMOSAURUS Marine creatures from the Cretaceous Age. Fossilized remains of which, estimated as being 80 million years old, were found in the late 1980s in the sedimentary rocks of both the Puntledge and the Brown Rivers, near Courtenay.

ERODED Action by the forces of nature which wears away the earth's erosion surface.

FORAMINIFORS Marine plankton.

FOSSIL Organic remains from a former geological age buried by natural processes, and subsequently permanently preserved.

FRENCH, JESSIE Granddaughter of George Ford, one of the island's earliest settlers. Jessie, who died in 1986 at the age of eighty, was a fund of island knowledge, both practical and mystical.

GABRIOLA FORMATION The last cycle of disposition of Cretaceous sediment on Hornby. Comprised mainly of conglomerate rock, it overlies the Spray Formation on the St. John's Point Peninsular.

GASTROPODS Mollusk class that includes the snail.

GEOFFREY FORMATION Third cycle of deposition of Cretaceous sediment on Hornby. Laid down in a river mouth environment, it comprises mainly conglomerate rock. It appears along the Goose Spit between the Shingle Spit and Ford's Cove, over on Downes Point, and again between Collishaw Point and Galleon Beach.

GINGKO TREE Large ancient gymnospermous tree, with fan-shaped leaves and fleshy fruit; its descendant still to be found as an ornamental tree in China.

GNEISSES Metamorphic rocks, generally made up of bands that differ in color and composition, some bands being rich in feldspar and quartz, others in hornblend or mica.

HOT SPOTS Plumes of hot rock welling up from deep in the mantle. They are isolated areas of geological activity, and may occur in the middle of a mobile plate or at the midocean ridges where two plates spread apart. They cause volcanoes, earthquakes, and keep the ocean plates moving. The plumes are thought to be static and crustal plates drift over them.

IGNEOUS ROCK Formed from magma that rises from the earth's interior. Magma that solidifies before reaching the surface forms intrusive rocks like dolerite, gabbro and granite. Magma that solidifies after reaching the surface forms volcanic rocks like basalt, obsidian and pumice.

IRIDIUM A metal, rare on planet Earth, though sometimes found in the effluence of volcanoes; believed to be relatively abundant in terrestial bodies.

JUAN DE FUCA PLATE Small oceanic plate that came into being about 50 million years ago, when a new ocean ridge was created adjacent to the west coast of Vancouver Island.

KARMUTSEN FORMATION The oldest part of the "Vancouver Group". The name given to most of the rocks underlying the Cretaceous sediment of Vancouver Island and the Queen Charlotte Islands. It is the thickest and most widespread formation on Vancouver Island. Relics from it show on the Beaufort Range, at Lasqueti, again on Texada and up the Forbidden Plateau to Buttle Lake and across to Quadra Island. Karmutsen rocks have been faulted, crumpled, eroded, and in some places the age of Karmutsen rock is determined by the Early Permian age of underlying Buttle Lake Limestone and the early Late Triassic age of overlying Quatsino Limestone. A thin band of intervolcanic limestone, interbedded in the Karmutsen, several hundred feet below the top, carries early Late Triassic fossils.

LAVA When magma reaches the earth's surface, it is termed lava. Rocks that are formed from lava are termed volcanic rocks.

LARAMIDE OROGANY Geological term used for the tectonic action that created the Rocky Mountain Cordillera.

LEMUR Nocturnal mammal, with foxlike face and woolly fur, that lives in trees, is related to the monkey, and today is to be found mainly in Madagascar.

LITHIFACTION The process which results in the formation of a massive rock from a loose sediment.

LITHIFYING Hardening sediment into rock.

MARCASITE Crystalline iron pyrites.

MAGMA Molten material within or beneath the earth's crust from which igneous rock is formed.

MARSUPIAL A mammal, such as an opossum, or a kangaroo, that carries its young in a pouch.

METAMORPHIZED Rocks transformed and crystallized by heat and pressure. In this process, the minerals of the original rock either reform into larger crystals or react together to form new minerals. Rocks produced have a layered texture called foliation, as in slate. Metamorphic rocks include, besides slate, which is produced from clay at a low temperature, schist and gneiss, formed at a higher temperature, and under greater pressure, and marble that is formed from limestone.

MIOCENE An epoch in the Tertiary period, 12 million to 25 million years ago; characterized by the presence of grazing mammals.

MOSASAUR Carnivorous, marine lizard from the Cretaceous period.

MUDSTONE A clay-like rock of nearly uniform texture throughout with little or no lamination.

MULTITUBERCULATES Early fruit-eating mammals. Rodent like, and varying in size from that of a mouse to a terrier.

NORTHUMBERLAND The oldest formation of Cretaceous sedimentary rock on Formation Hornby. Composed mainly of mudstone-turbidite, laid down in a shallow sea environment, it is found outcropping in Shingle Spit and Collishaw Point. There is also some to be seen slightly inland between Ford's Cove and Heron Rocks.

OCEANIC PLATES Oceanic plates are created and propelled by hot magma bubbling from the earth's core, through mid-ocean ridges.They are conveyor belts that, along with hot spots keep the land masses of the globe constantly on the move. They are thinner than continental plates, varying from twenty-five to one hundred kilometres in depth and, as they are constantly being recycled are never more than two million years old. As they move under the sea, they gather sediment, traces of which are revealed when subduction takes place. As the plate plunges back into the mantle of the earth, parts of the oceanic crust are uplifted, and left behind to form, along with molten magma from the earth's mantle, new mountain ranges. See "Tectonic Plates".

OLIGOCENE Epoch in the Tertiary period from 25 million years ago to 40 million years ago.

OYSTERCATCHER Long red-billed wading bird, with black plumage and conspicuous yellow legs.

PANGAEA Name given by geologists to the super continent, composed of all the major continental plates, believed to have existed 300 million to 400 million years ago.

PANTHALLASAH Name given to ancient sea, located 350 million years ago in the latitude of the present South Pacific.

PELECYPODS Mollusk with bivalve shell, belong to the class that includes oysters, clams, mussels, scallops, etc.

PLANKTON The minute, floating organisms of plant and animal origin, usually invisible to the naked eye, which inhabit the waters of oceans and seas, and form the food of many fishes and other creatures.

POLARITY Manifestation of two opposite tendencies; the negative and positive pull of iron magnets to the magnetic poles. Polarity is said to change when the compass instead of pointing north, points south. It is thought that the earth has an iron nickel core which, as it rotates, induces magnetic forces. These extend beyond the surface of earth to form the magnetosphere. The earth's magnetic field periodically undergoes abrupt reversals in direction of polarity accompanied by changes in intensity. The magnetic north pole migrates toward the geographic south and the magnetic south pole migrates toward the geographic north pole. When this happens the compass points south, instead of north. These reversals are recorded in igneous rocks. There is usually a few thousand years between the establishment of the reversal when the earth's experiences no magnetic field at all. In the last 10 million years reversals have occurred approximately once every million years.

RIDGE A long narrow elevation of land; a chain of mountains or hills.

RIFT A fault; the valley along the edge of a fault.

SANDSTONE Sedimentary rock composed of small grains, usually quartz and feldspar cemented together.

SCHIST Fine-grained metamorphic rock with component minerals arranged in more or less parallel layers. Splits in thin, irregular plates.

SEA LEVELS Three factors affect changing sea levels; eustasy, isotasy and tectonic plate movements. Eustasy means a world-wide change in sea levels, usually due to melting of glaciers. Isotasy is the tendency of the earth's crust to maintain a state of equilibrium. Continental plates float on a denser substratum. If weight is removed from a mountain range by erosion, the range will rise slightly to compensate for the loss of land above the water surface. When it is the weight of a glacier that is removed the land rises quickly. The land may be submerged again as eustasy, from the world-wide melting of glaciers, takes place. At the end of the last glacial period, 13,000 to 11,500 years ago, a see-sawing between the land and water levels that took place in this area. The pattern is not consistent, and research is still being done on the causes for this.

SEDIMENTARY ROCKS Sand, silt or pebbles, eroded from other forms of rock, carried elsewhere by wind or water, and then cemented, by chemical action and/or the weight of more detritus. They have a layered structure and contain all the world's oil and coal and all its fossils.

SHALE Very fine grained laminated sedimentary rock consisting of consolidated mud or clay.

SICKER GROUP Basaltic flows. Breccias, tuffs, graywacke, sandstone, chert, shale, limestone, dating back 380 to 280 million years. Sicker Rocks are to be found in discontinuous bands south centre of Saltspring Island to Hesquiet Inlet on the West Coast of Vancouver Island; from Cowichan to Horne Lake; in the Buttle Lake area and across from Nanoose to Texada. In the Buttle Lake limestone are found Permian fossils.

SPRAY FORMATION The fourth cycle of deposition of Cretaceous sediment on Hornby Island. Laid down in a delta and marine environment, it comprises sandstone, mudstone, and shale. The Spray Formation can be seen on Sandpiper Beach, between Downes and Dunlop Points, and over on the far side of Tribune Bay. On the St. John's Point Peninsular it underlies the Gabriola Formation. On the northern beach it can be seen for about half a mile west, and about one quarter mile east of Tralee Point. It can also be seen from Collishaw Point east to Galleon Beach.

STIKINE TERRANE A tectonic block of exotic origin, now comprising much of central and northwestern British Columbia.

SUBDUCTING Term used when an oceanic plate goes under another plate, and is assimilated into the mantle. See Tectonic Plates.

TECTONIC PLATES The continents and oceans of the world rest on a series of giant interlocking plates; some of the plates are oceanic and some continental. They float on an underlying sea of semi-molten rock, and are constantly on the move. It is as though the globe was covered with a shifting jigsaw puzzle. The mechanics of plate motion are the subject of great debate. Ocean ridges are underwater mountain chains through which lava is constantly erupting and forming new basaltic oceanic crust. The end of the plate furthest from the ridge, meeting an obstacle such as another oceanic plate or a continental plate, plunges back into the centre of the earth melting and releasing great heat as it goes. As this happens, tensions in the earth's crust cause earthquakes, magma seeking to escape activates volcanoes, and an upward buckling of the oceanic crust creates fresh mountain ranges. (See Continental Plates and Subducting.) The acceptance by earth scientists of the concept of tectonic plate action has not only answered many, up to now, enigmatic questions, (why for example there are ancient marine fossils in the Rocky Mountain range), but has expanded human understanding as to the cyclical interdependence of the whole global machine; made it aware of the total inter-connected ecosystem—the recurring pattern of death and rebirth—of the entire planet earth.

TERRANE Term used to describe pieces of land, with varying origins, that journey independently on the movement of oceanic plates.

TERTIARY Period following the Cretaceous, 10 million to 65 million years ago, characterized by the development of mammals, and the arrival of grasslands.

VOLCANOES A volcano is a vent or fissure in the earth's crust through which molten magma, hot gases, and other fluids escape to the surface of the land, or on to the bottom of the sea. Volcanoes are an essential part of the earth's ecosystem. Without volcanoes there would be no creation of new oceanic crust through seafloor spreading, no subduction (the dragging down of older crust into the mantle to be melted and recycled), no mountain building, and in conjunction with this no erosion and no sedimentation.

WRANGELLIA The name given by geologists to land often known in Canada as the Insular Belt; the word is taken from the Wrangell Mountains in Alaska.

Glossary for Part II

THE FIRST PEOPLE

CLOTHING

The use of skins for clothing in a world of constant rain was not practical. Instead, the specially treated inner bark of cedar or lengths of tule or sedge, woven into cloth, was fairly waterproof and light to wear. Women wore skirts of cedar bark. Young and middle aged men wore a belt to which was attached a free hanging cover for their groin made of cedar bark fibre or skin. When working, they drew it up through their legs and fastened it at the back. Old men wore nothing at all. For an outer covering, both sexes wore a cape, or slung a robe over one shoulder, fastening it with a belt. Hats were mushroom shaped of woven basketry. Most of the time the Pentlatch went bare footed but would wear skin moccasins when walking in prickly areas, such as among the cactus on St. John's Point.

COAST SALISH

In prehistoric times, people who occupied land from Quadra Island in the north to Washington state in the south. They were an unwarlike, industrious, semi-nomadic people, divided up, not into tribes or bands, but into families and extended families. Each family spent the months of winter in big houses in winter villages, breaking away in the spring into smaller family units. Like flocks of birds that come together, and then part, they flitted within the territories that they "owned", camping, fishing, hunting, gathering wherever and whenever resources were ripe for harvesting. Each geographical area of families, and even each family, differed from the next in dialect and in details of living and ritual. The people in the north could not understand the people in the south. However, they shared the same language roots, the same patterns of economic and social survival, and, in some cases, the same myths. One myth told that they had always been here. They all breathed the same shamanic world of the supernatural. See also Social Structure and Economic Structure.

ECONOMIC STRUCTURE

Following traditions going back thousands of years, the Coast Salish people evolved and continued to evolve a many- dimensional network of getting and sharing, gathering and preserving, giving and taking. Without money tokens or direct trading, and without domination of others, they created a system that made sure that no surplus food went to waste and that bounty when it occurred was shared both by close relatives, and by families related through marriage. This was not entirely altruistic, although generosity was a respected quality and an indication of personal power. The giving to people outside the immediate family was a way of storing "capital", and an insurance against bad times. All those who received bounty were expected to return food of an equivalent value when a surplus of food happened for them. The system was full of such balances, all of which were closely tied in with social rituals. For example, to mark the arrival of a gift of food from relatives by marriage, both families were invited by the receiving party to a feast.

EDUCATION AND DISCIPLINE

This was by word of mouth and by example. Successful people were those who listened to their grandfather's or grandmother's words. Knowledge was power, and acquisition of special knowledge perpetuated a person's position within the family. There was unrelenting pressure from the elders in the family—both by example and through stories—for every one within the family to be brave, industrious, generous and to follow all the rules of ritual and behaviour. This was not only for each individual's own sake but also for the well-being and respect of all the family and extended family. Physical toughness and cleanliness was also part of the training. Children bathed everyday regardless of how cold the water might be. Even more demanding than the eye of the elders was the presence of the world of the spirit. Children were brought up to fear the consequences if they did not keep to rituals and respect taboos. It was not spelled out what would happen if they did not do this, except that there would be a loss in spirit power and with this, a loss in special skills and general ability.

DEATH BELIEFS There was no one place to which departed spirits went, though some places were believed to hold more spirit power than others. Ghosts were shadows that hung about near burial grounds. They could steal souls away. Downes Point Caves, The people made a distinction between soul and body. The soul was mind or consciousness, free to leave the body during a person's life. After death it could come back in the body of an owl, or another creature, or as one of his or her descendants. Breath was the vitality in the body. If breath left the body it died. Breath was spirit power within a person, as opposed to external spirit power. Breath was also the essence of a person's personality. It might or might not be reincarnated.

DEATH After death occurred in a family, protective rituals took place in charge of an older person. The family members had to purify themselves by washing in salt water. The house had to be exorcised with the scented steam of an odorous plant stirred in water heated by hot rocks, and by the brushing of burning boughs, four times, into the corners and through the overhead space. The widow had to live for sixteen days outside the house, and could not pass through any door. An attendant would bring her food, and see that every fourth day she had a bath. On the sixteenth day, all the clothes that she was wearing were hidden away in an old stump.

ECOSYSTEMS Within the period 500 A.D. to 1790 A.D., while seasonal and cyclical patterns did not alter appreciably, the size and numbers of people visiting the island did. The size of population fluctuated in the same way that the place and quality of food gathering fluctuated. It was never exactly the same twice. The picking areas would vary from year to year. Climate changes, a series of very wet or very dry years, altered vegetation patterns; earthquakes and high tides eroded shore lines and cliff; and fires, started by lightening or negligence, would have on occasion swept a stand of first growth timber back to charred remains and on to the birth of a new cycle of growth. Uses were found by the people for everything offered within each cycle of renewal. Whether nettles, thistles, and horsetails; or berry bushes, and young alder; or maples, crabapples, cherry and oceanspray, or young fir, hemlock, spruce, and cedar; each phase had its uses. Eventually the old forest came back.

EUCLATAWS Term used by early settlers for some of the Kwakiutl people.

FEARS While it may seem that the people who came to this island lived in an almost ideal world of harmony and plenty, this was not entirely so. There was the fear of censure from others, both from within and outside the family, and there was fear of their Salish neighbours, who spoke in different dialects and who might start a night raid on the slightest provocation.

Above all, there was the fear of the supernatural should actions be not pleasing to the spirits or should the people, through incorrect behaviour, have left themselves open to malevolent spirit power.

GULF OF GEORGIA PHASE Also known to archaeologists as the "Developed Coast Salish Phase", is the name given to artifacts found dating back to the Phase period 500 A.D.to 1800 A.D. The phase is marked by some changes in the fashioning of tools, by a decline in the stone sculpture of the preceding Marpole phase, and by the advent of a weaving complex with carved spindle whorls and elaborate blanket pins. See Bibliography.

HAIDA West coast Indian people inhabiting the Queen Charlotte Islands.

HAMMERSTONE According to Hilary Stewart, "a multi-purpose tool with an infinite range of uses". The hammerstone in this book appears, from wear, to have been used for banging in wedges to split wood. It is a cobble of amygdaloidal lava. This lava is found in the rocks of the Beaufort Range, on Vancouver Island, and dates to the final flows of the Karmutsen volcanic eruptions. The cobble has been smoothed and rounded by water friction. It could have been cracked conglomerate, or split from the Beaufort Range rock in the Pleistocene with the aid of the glaciers.

KWAKIUTL A west coast Indian people living on the northern part of the east coast of Vancouver Island and on the adjacent mainland.

LITHIC STAGE Name given by archaeologists for the period 9,000 B.P. to 4,000 B.P. The oldest lithic stage assemblages so far identified, 9,000 years—are probably about 2,000 years younger culmination of a considerable span of prior adaptation to the local region. The reason for there not being traces of settlements earlier than 9,000 years ago could be on account of the changing sea levels. Similarly, the apparent hiatus between the Lithic Stage and what are known as the development cultures—such as Locarno Beach and Marpole—is again because of sea levels changing. The early lithic assemblages seem primitive, possibly because in them there are only stone artifacts. An absence of shell in a midden could have also led to the disappearance of all evidence of bone. The presence of shell counters the acidity of the soil. See Bibliography.

MARPOLE PHASE Also known as the Middle Development Phase. Covered the period 400 B.C. to 500 A.D. Stone sculpture attained a peak of quality and excellence; basketry, cordage, fish nets, bent wood boxes, decorated bowls, canoe paddles, wood wedges, clubs and masses of wood chips and splinters, have been found in waterlogged sites, indicating a prosperous society. Burial inclusions and the practice of artificial skull deformation suggest a society conscious of class. Marpole, the site that has given this phase its name, is now to be found under the parking lot of the Fraser Arms Hotel, Vancouver, British Columbia.

MARRIAGE Socially and economically one of the most important threads in the Coast Salish tapestry. Girls and boys, particularly those born into the privileged class, knew that they would need to marry outside their extended family. Even the marriage of second cousins was frowned on. Marriages cemented friendships between families, and made for peaceful relationships; marriages, through the rituals of exchange spread around the availability of goods and balanced the economy. The bethrothal was formal. The betrothed sat on a pile of cedar bark blankets, while the elders and people of both families feasted, and gave effusive speeches as to the qualities of the bethrothed pair. The boy's family gave food, blankets, and wealth of all kinds. All the girl's parents gave to begin with were her personal household utensils, but over the years, gifts were given back and forth. In addition to material goods, permission might be given to use a family crest, along with the songs and dances that went with it. This would be on condition that such privileges were handed on to the children. It was not always the girl who had to leave home. Sometimes a man lived with his wife's family.

MIDDEN Refuse dump. Evidence of habitation by people shown initially by dark shell flecked soil. Besides shells there may be found bones, human and otherwise, bone and stone artifacts, and the remains of fireplaces. Meticulously excavated, layer by layer, by archaeologists, the pattern of living and historical content of the first people on the island can be reconstructed. Haphazard digging—by amateurs—although artifacts may be found, is liable to destroy sequential evidence.

MOUNT MAZAMA A volcano whose massive eruptions 6,600 years ago blanketed ash over an area of 350,000 square miles. Charred wood found in and beneath pumiceous sheets, radiocarbon dated, gives geologists the date of this eruption, and gives archaeologists, finding layers of this ash in Fraser Valley digs, an indication of the time slot they are excavating. The core of the volcano can be seen today in Crater Lake, south west Oregon.

NAMES Names had privileges attached, and were often symbols. Not until they were ten months old were babies given their official names. This was a sign of privilege. Names were also given later in life at special ceremonies.

ORNAMENTS On ceremonial occasions, people of importance wore bone and shell pieces in their ears and noses, and strung many strands of shell or bone beads around their necks. Sometimes the Salish rubbed their entire bodies with deer or seal grease as a protection against the cold. The grease also soothed dry skin, and stopped mosquitos from biting. On ceremonial occasions, and when they needed to look fierce, the Salish greased their hair, drenched it in red ochre and painted their faces with ochre and charcoal black.

PENTLATCH Name given to the people living in the vicinity of Hornby Island prior to the arrival of the white man, and prior to the movement south of the Comox people. Sickness and raiding led to the virtual extinction of these people. By 1862 the Pentlatch (pronounced Puntledge), were reduced to one big house near the mouth of the Puntledge River. Homer Barnett, in his book *The Coast Salish,* says "The Pentlatch proper did not extend as far south as Qualicum Bay . . . but their language was spoken down to Nanoose Bay . . . their habitations were certainly along Comox Harbour and the mouth of the Puntledge River which flows into Comox Harbour. Southward from Union Bay to Deep Bay, lived their linguistic relatives the s:uckcan. The saaLam ranged from that region to Englishman River."

PETROGLYPH Prehistoric rock carvings of which the origin and meaning, to a great extent, is still unknown. They are physical manifestations of myth and meaning now lost to us. Unless a carving has been uncovered by soil containing remains of carbon, it is difficult to establish a date. Some obviously post-date the arrival of the white man, i.e. the carving of the Hudson Bay steamship "The Beaver", at Clo-oose on the West Coast of Vancouver Island. Some of the petroglyphs on Hornby are shamanic, with strong spiritual associations, and some are symbols of spirit helpers. It is thought that individual fishermen and sea-mammal hunters re-pecked and re-ground their symbol, while singing their spirit song. This enabled the spirit in the carving to come out of the rock and assist in the fishing or the hunt. Some petroglyphs are directional signs. A petroglyph at Whaling Station Bay—a person with fingers of one hand spread out and one leg raised—Mrs. Mary Clifton of Comox believed, is a sign that fresh water is close. The number of fingers indicates the number of canoe lengths away and the knee raised up indicates that it is necessary to climb. One conjecture about the petroglyphs near Tralee Point is that most of them date from the nineteenth century. Carvings older than this, unless constantly pecked and re-ground, as they may have been, would by now have been quite eroded by the tides. The collapse of society during the nineteenth century would have accelerated ritual. Rock carving is one way the people might have felt they could regain some of the ancient power of the spirit.

POPULATION CONTROL There was little infanticide among the people, but there were women who knew how to procure abortions. Continence was practiced through ritual. This, intentionally or not, kept the population within resource limits. When there were too many children in a family, an uncle or cousin would take one or two from the family and rear them.

POTLATCHES Potlatches gave more prestige than feasts. Feasts extended friendships, balanced food supplies, banked food capital. Potlatches, although feasting accompanied them, dealt with 'real wealth'. Invitations were sent further afield, and to non-relatives, as well as relatives (although speeches were made embracing everybody present as a relative). It was not just the headman of the house or village who gave the potlatch, though he was the funnel through which it happened; the whole family was responsible. It might take years for enough goods to be accumulated to make a potlatch possible. Goods acquired at one potlatch might be given away at the next. It was usual to try to give back more than had been given. Potlatches built up capital and prestige for the giver, and kept wealth circulating. They were also used as an occasion to establish "claims". As there were no written records, statements concerning the giving of privileges, the awarding of honorific names, and the transfer of property needed to be witnessed. In addition to the public announcement, special witnesses were paid to corroborate the statements should the need arise at any time.

RAIDS

During the winter, raids took place on villages—raids to avenge a slight, acquire booty, or relieve winter boredom. Before the nineteenth century, raids did not normally happen during the spring to fall gathering, hunting time. Everyone was too busy.

RITES OF PASSAGE

The people believed that during all rites of passage, during times of bodily change—birth, puberty, death—a person was especially vulnerable to spirit power. To avoid danger—for both the person in question and those associated with him or her, from malevolent spirits—certain taboos had to be observed at such times and certain rites followed.

SLAVES

Slaves, usually also Salish, were war captives and though absorbed into the family, and not treated unkindly, were deprived of any privileges. Their blood was considered tainted for several generations.

SOCIAL STRUCTURE

The Salish Society was one that survived by mutual inter-dependence. To a great extent it was egalitarian, and cooperative. Within each extended family there was a recognizable hierarchy. There were those who inherited, or through having special skills acquired, property (property being not only land rights, and possessions, but also privileges, ritual magic, special songs); from amongst these privileged people, headmen were chosen. Between them and the people without any status were those whose social position was fluid; many were specialists, in magic, in crafts or in healing. Those at the bottom of the scale, who had nothing, might be there by virtue of their birth —they may be the descendants of a slave—or they might be there by having suffered in some way from "loss of power".

SPIRIT POWER The world of the spirit and their relationship to it was of great importance to the Salish. Success in everything was equated to spirit power. In the dark days of winter, cooped up in their winter villages, the Coast Salish people's creativity was channelled not, as with other West Coast people, into totems and objects of status, but into developing their understanding and relationship with the world of the spirit. Every tool and utensil, every basket, blanket, and spindle, that they made, they endeavoured to imbue with beauty and with spirit power. Within each extended family there were certain people who specialized in spirit power; herbalists, ritualists, clairvoyants, and magicians as well as shamans. Power came by both revelation and through knowledge passed on by the elders—knowledge handed down through the centuries. Each person as he or she passed into puberty came to know their special spirit helper or guardian spirit. See Rites of Passage; Spirit Quest.

SPIRIT QUEST Puberty was a time of change when a child going into adulthood, and was considered extremely vulnerable to spirit influences. It was, therefore, that special taboos and rituals were followed at this time. At this time too, through a special quest, or by other means producing vision, the pubescent child's guardian spirit was revealed. The vision also gave a special song and a special dance. This was very personal and only revealed to others at the winter dances. The boy's ordeals were not supposed to involve physical suffering. They were primarily exercises overcoming fear. Fear of ghosts and other dangerous beings lurking in the dark was very real. At the end of a quest, with the knowledge through vision of whom his guardian spirit was, whether a wolf, or a whale, or some supernatural creature, came a spirit song and a spirit dance. These secrets the boy kept to himself. Only in the winter dances, when young and old had a chance to dance their dance and sing their song, was an intimation of the vision revealed. A girl, to gain her vision, song and dance, had to find ways other than deprivation in the woods. It was thought more important for girls to make good marriages—by being industrious, demure and attractive—than to have occupationally helpful spirits.

TLINGIT A native people inhabiting part of northern coastal British Columbia and Alaska.

TULE Bulrush. Used by people to weave mats and capes.

WINTER HOUSES The nucleus of the Coast Salish family. The storehouse and generating centre for all the threads of Coast Salish life. Into the house was brought, in the fall, all the raw materials, of stone, bone and fibre, needed for the creation of non-perishable goods, and all the dried and preserved food collected during the summer. Out of the house in the spring came all the newly made tools, nets, baskets, fish hooks, and mauls ready for use during the longer season of camping, hunting, fishing and gathering. Within the winter houses instruction took place. Myths were recounted, and spirit dances displayed, dances that meant a chance for individual recognition as well as for communal feasting. There were usually three to four families in each house, perhaps one or two houses to a village.

WINTER VILLAGES Usually situated at the mouth of a river away from winter storms and hidden from raiders. It was from the winter village, through potlatches and feasts, that relationships with neighbouring villages were strengthened, economies balanced, marriages arranged and gifts or inheritances announced.

Bibliography

Foreword

Ludvigsen, R. and Beard, G., *West Coast Fossils: A guide to the Ancient Life of Vancouver Island.* Madeira Park, B.C.: Harbour Publishing, 1984.

Ludvigsen, R. and Beard, G., *West Coast Fossils: A guide to the Ancient Life of Vancouver Island* (Revised Edition). Madeira Park, B.C.: Harbour Publishing, 1994.

Chapters 2 and 3

Burchfiel, B.C. "The Continental Crust." *Scientific American,* vol. 249, no. 3 (1983): 130.

Dietz, R.S. and Holden, J.C. "Reconstruction of Pangaea: breakup and dispersion of continents, Permian to Present." *Journal of Geophysical Research,* 75 (1977): 4939-956.

Francheteau, J. "The Oceanic Crust." *Scientific American,* vol. 249, no. 3 (1983): 114.

Irving E., Monger J.H.W. and Yole R.W. "New Paleomagnetic Evidence for Displaced Terranes in British Columbia." *Geological Association of Canada Special Paper,* no. 20 (1980): 441-456.

Jones, David L., Cox, Allan, Coney, Peter, and Beck, Myrl. "The Growth of Western North America." *Scientific American,* vol. 247 (1982): 70-84.

Monger, J.W.H., Souther, J.G., and Gabrielse, H. "Evolution of the Canadian Cordillera: A plate tectonic model." *American Journal of Science,* vol. 272 (1972): 277-602.

Monger, J.W., "Evolution of the Cordillera." *Geos, Energy, Mines and Resources, Canada,* Fall (1978): 5.

Monger, J.W.H. "Tectonic accretion and the origin of two major metamorphic and plutonic welts in the Canadian Cordillera." *Geology,* vol. 10 (1978): 70-75.

Muller, J.E. "Evolution of the Pacific Margin, Vancouver Island and Adjacent Regions." *Canadian Journal of Earth Sciences,* no. 14 (1977): 2062-85.

Muller, J.E. and Carson, D.J.T. "Geology and Mineral Deposits of Alberni Map Area, British Columbia (92F)." *Geological Survey of Canada,* paper 68-50 (1969): 52.

Siever, Raymond. "The Dynamic Earth." *Scientific American,* vol. 249, no. 3 (1983): 46.

Vink, G.E., Morgan, W.J., and Vogt, P.R. "The Earth's Hot Spots." *Scientific American,* vol. 252, no. 4 (April 1985): 50.

Wertenberger, William. *The Floor of the Sea.* Toronto: Little, Brown & Company, 1984.

Yole, R.W. "An Early Permian Fauna from Vancouver Island, British Columbia." *Bulletin of Canadian Petroleum Geology,* vol. 2, no. 2. (1963):138-49.

Yorath, C.J. and Cameron, B.E.B. "Oil off the West Coast." *GEOS,* vol. 2, no. 2 (1982).

Chapters 4, 5 and 6

Bell, W.A. "Flora of Upper Cretaceous Nanaimo Group, Vancouver Island." *Geological Survey of Canada Memoir* 293, (1957): 77.

Fiske, D.A. *Stratigraphy, Sedimentology and Structure of the Late Cretaceous Nanaimo Group, Hornby Island, British Columbia.* Corvallis: Oregon State University, 1977.

Halstead, L.B. *The Search for the Past.* Toronto: Doubleday, 1982.

Muller, J.E. and Jeletsky. "Geology of the Upper Cretaceous Nanaimo Group Vancouver Island and Gulf Islands, British Columbia." *Geological Survey of Canada,* paper 69-25 (1970): 77.

Rouse, G.E., Mathews, W.H., and Blunden, R.H. "The Lions Gate Member: A New Late Cretaceous Sedimentary Subdivision in the Vancouver Area of British Columbia." *Canadian Journal of Earth Sciences,* vol. 12, no. 3, (1975): 464-71.

Usher, J.L. "Ammonite Faunas of the Upper Cretaceous Rocks of Vancouver Island, British Columbia." *Geological Survey of Canada,* bulletin 21, (1952).

Ward, Peter, "The Extinction of the Ammonites." *Scientific American,* vol. 249, no. 4 (1983): 136.

Chapter 7

Rouse, Glenn E., Hopkins, W.S., and Piel, K.M. "Palynology of some Late Cretaceous and Early Tertiary Deposits in British Columbia and Adjacent Alberta." *Geological Society of America,* Special Paper 127 (1971): 213-246.

Rouse, Glenn E. *Paleogene Palynomorph ranges in Western and Northern Canada.* 1977.

Thomson, Richard E. "Oceanography of the British Columbia Coast." *Fish and Aquatic Sciences* 56 (1981): 29l.

Thornton, Harry. "Icebound Eden." *Equinox,* no. 27 (1986): 72.

Chapter 8

Alley, Neville F. "Middle Wisconsin Stratigraphy and Climatic Reconstruction southern Vancouver Island, British Columbia." *Quaternary Research,* vol. 2, no. 2, (1979): 213-237.

Claque, J.J. "Late Quaternary Sea Level Fluctuations, Pacific Coast of Canada and adjacent areas." *Geological Survey of Canada*, paper 75-1 C. (1975): 17-21.

Fyles, J.G. "Surficial geology of Horne Lake and Parksville map-areas, Vancouver Island, British Columbia." *Geological Survey of Canada Memoir* 318 (1963).

Mathews, W.H., Fyles, J.G., and Nasmith, H.W. "Postglacial crustal movements in southwestern British Columbia and adjacent Washington State." *Canadian Journal of Earth Sciences*, vol. 7, no 2 (1970): 690-702.

Stalker, Archie. "Ice age bones—a clue." *GEOS*, vol 13, no. 2, (1984): 11.

Chapter 9

Borden, Charles E. "Origins and Development of Early Northwest Culture to about 3,000 B.C. National Museum of Man, Mercury Series." *Archaelogical Survey of Canada*, paper 45, (1975): 137.

Bunyan, D.E. "Pursuing the Past. A General Account of British Columbia's Archaeology." *UBC Museum of Anthropology*, museum note no. 4, 1978.

Burley, David V. Marpole. "Anthropological Reconstructions of a Prehistoric North West Coast Culture Type." *Department of Archaeology, Simon Fraser University* 8. Vancouver: Simon Fraser University, 1980.

Bernick, Kathryn. "A Site Catchment Analysis of the Little Qualicum River Site DiSc l: A Wet Site on the East Coast of Vancouver Island, British Columbia." Ottawa: National Museum of Man, 1983.

Capes, K.H. "Contributions to the Prehistory of Vancouver Island." Occasional Paper 15. Pocatello: Idaho State University Museum, 1964.

Capes, K.H. "Archaeological Investigations of the Millard Creek Site, Vancouver Island, British Columbia." *Syesis*, vol. 10 (1977) 47-84.

Claque, J.J. "Late Quaternary Geology and Geochronology of British Columbia. Part l. Radio Carbon Dates." *Geological Survey of Canada*, paper 80 (1980): 13.

Drucker, P. *Indians of the Northwest Coast.* Garden City: Natural History Press, 1965.

Drucker, P. *Cultures of the North Pacific Coast.* San Francisco: Chandler Publishing Company, 1965.

Duff, Wilson. *Images in Stone, B.C. Thirty Centuries of Northwest Coast Indian Sculpture.* Sannichton: Hancock House, 1975.

Fladmark, K.R. "An Introduction to the Prehistory of British Columbia." *Canadian Journal of Archaeology*, no. 6 (1982): 95-156.

Hebda, R.J. and Rouse, G.E. "Palynology of two Holocene cores from the Hesquiat Pensinsular, Vancouver Island, British Columbia." *Syesis*, vol. 12 (1979): 121-29.

Heusser, Linda E. "Palynology and Paleoecology of post glacial sediments in an anoxic basin. Saanich Inlet, British Columbia." Lamont Doherty Geological Observatory, Columbia University, Palisades, N.Y.

Keen, Sharon Denise. "The Growth of Rings of Clam Shells from two Pentlatch Middens as indicators of Seasonal Gathering." *Heritage Conservation Branch* no. 3.

Mathewes, Rolf W. and Heusser, Linda E. "A 12,000 year Palynological Record of Temperature and Precipitation Trends in South Western British Columbia." *Canadian Journal of Botany*, vol. 59, no. 5 (1981): 707-10.

Mitchell, Donald H. "Archaeology of the Gulf of Georgia area. A natural area and its culture types." *Syesis*, vol. 4, supplement l (1971): 228.

Mitchell, Donald H. and Bernick, Kathryn. *Preliminary Report of Salvage Excavations at Dksf 26, Courtenay.* Victoria: University of Victoria,1981.

Monks, Gregory G. Excavations at the Deep Bay Site (DiSe7) Vancouver Island. Annual Report, Archaeological Advisory Board, 1977.

Whitlam, Robert. *Salvage Excavations at Buckley Bay Site (Djsf 13) and at Tsable River Site (Djsf 14) Preliminary Report, Department of Anthropology.* Victoria: University of Victoria, 1974.

Chapters 10 to 18

Amoss, Pamela. *Coast Salish Dancing. The Survival of an Ancestral Religion.* Seattle: University of Washington, 1978.

Barnett, Homer G. *The Coast Salish of British Columbia.* Eugene: University of Oregon Press,1955.

British Columbia Heritage Series l. *Our Native Peoples*, vol. 2, Coast Salish. Victoria: British Columbia Provincial Museum, 1965.

Eliade, Mircea. *Shamanism, Archaic Techniques of Ecstasy.* Trans. by Willard R. Trask. Princeton: Princeton University Press, 1964.

Hill, Beth and Ray. *Indian Petroglyphs of the Pacific Northwest.* Surrey: Hancock House, 1974.

Jenness, Diamond. "The Faith of a Coast Salish Indian." Edited by Wilson Duff. *Anthropology in British Columbia*, Memoir no. 3, Victoria: British Columbia Provincial Museum, 1955.

Jilek, Wolfgang G. *Indian Healing. Shamanic Ceremonialism in Pacific Northwest Today.* Surrey: Hancock House, 1982.

Kennedy, Dorothy and Bouchard, Randy. *Sliammon Life, Sliammon Lands.* Vancouver: Talon Books, 1983.

Levi-Strauss, Claude. *The Way of the Masks.* Trans. by Sylvia Modelski. Seattle: University of Washington Press, 1982.

Levi-Strauss, Claude. *Myth and Meaning.* Toronto: University of Toronto Press, 1978.

Maud, Ralph, Ed. *The Salish People, The Local Contributions of Charles Hill-Tout. Volume IV: The Sechelt and and the South Eastern Tribes of Vancouver Island.* Vancouver: Talon Books, 1978.

Robinson, Sarah A. *Spirit Dancing Among the Salish Indians, Vancouver Island, British Columbia.* Chicago: University of Chicago, 1963.

Stewart, Hilary. *Artifacts of Northwest Coast Indians.* Surrey: Hancock House, 1973.

Stewart, Hilary. *Indian Fishing. Early methods on the Northwest Coast.* Vancouver: Douglas and McIntyre, 1977.

Stewart, Hilary. *Cedar. Tree of Life to the Northwest Indians.* Vancouver: Douglas and McIntyre, 1984.

Suttles, Wayne. "Katzie Ethnographic Notes." *Anthrolopogy in British Columbia,* memoir no. 2. Victoria: British Columbia Provincial Museum, 1955.

Suttles, Wayne. "Affinal Ties, Subsistence and Prestige among the Coast Salish." *American Anthropologist,* vol. 62 (1960): 296-305.

Thompson, William Irwin. *The Time Falling Bodies Take to Light. Mythology, Sexuality and the Origins of Culture.* New York: St. Martins Press, 1981.

Turner, Nancy J. "Food Plants of British Columbia Indians. Part l." *British Columbia Handbook* no. 34. Victoria: British Columbia Provincial Museum, 1975.

Turner, Nancy J. "Plants in British Columbia, Indian Technology." *British Columbia Handbook* no. 38. Victoria: British Columbia Provincial Museum. 1978.

Whitlam, Robert. "Salvage Excavations at the Buckley Bay Site (DiSf 13) and the Tsable River Bridge Site (DiSf14). A Preliminary Report." Victoria: Department of Anthropology, University of Victoria, 1974.

Woodcock, George. *Indians of the Pacific Northwest.* Edmonton: Hurtig Publications, 1977.

Chapter 19

Colonist Newspaper Victoria. May 5th, p.3, Sept. 2nd, p. 3, 1863. Sept. 13th, p.3, 1867. Jan 13, p.3, 1869. All dates contain items regarding incidents on Hornby.

Duff, Wilson. "The Indian History of British Columbia, Vol. 1. The Impact of the White Man. Anthropology in British Columbia." Memoir no. 5, Victoria: Provincial Museum, 1964.

Kane, Paul. *Wanderings of an Artist Among the Indians of North America.* London: Longmans, 1859.

Mayne, R.C. *Four Years in British Columbia and Vancouver Island.* London: John Murray, 1862.

Newton, Norman. *Fire in the Raven's Nest.* Toronto: New Press, 1973.

Spradley, James P. *Guests Never Leave Hungry. The Autobiography of James Sewid, a Kwakiutl Indian.* Montréal: McGill-Queen's University Press, 1972.

RECENT CANADIAN NON-FICTION FROM NEWEST PRESS

Return to the Drum

Teaching Among the Dene in Canada's North

Miggs Wynne Morris

As a young teacher in 1965, Miggs Morris accepts a position to live and teach in a tiny, isolated community in what was known at the time as Fort Franklin in the Northwest Territories. As Morris shares her experiences of life with the people of Great Bear Lake—the Sahtuot'ine—her memories of these hard-working and proud people are interwoven with glimpses of their life from pre-European contact to the present.

ISBN 1-896300-31-6 • $24.95 CDN • $18.95 US

Watershed

Reflections on Water

Grant MacEwan

Water is the subject of the latest book by Alberta icon Grant MacEwan. MacEwan draws from his broad knowledge as an agriculturalist and his vast life experience to tell us "what every Canadian should know about water." These reflections are his love letters to water.

Watershed: Reflections on Water puts water in the context of its effect on our daily lives and our environment. MacEwan wants us to make tough decisions now in order to protect our water supply in the future.

ISBN 1-896300-35-9 • $19.95 CDN • $14.95 US